SpringerBriefs in Microbiology

More information about this series at http://www.springer.com/series/8911

Anshu Babbar

Streptococcal Superantigens

 Springer

Anshu Babbar
Medical Microbiology
Helmholtz Center for Infection Research
Braunschweig
Germany

ISSN 2191-5385   ISSN 2191-5393   (electronic)
SpringerBriefs in Microbiology
ISBN 978-3-319-22454-1   ISBN 978-3-319-22455-8   (eBook)
DOI 10.1007/978-3-319-22455-8

Library of Congress Control Number: 2015945609

Springer Cham Heidelberg New York Dordrecht London

Printed on acid-free paper

Springer International Publishing AG Switzerland is part of Springer Science+Business Media
(www.springer.com)

The role of the infinitely small is infinitely large

—Louis Pasteur

Preface

Streptococcal superantigens are a fascinating group of proteins that have evolved to modulate and thence evade host immune system. These molecules are exceptional in a way that they can stimulate adaptive immune response (activation of T-cells) in addition to the innate immune response. As a result, superantigens have ability to stimulate a large number of immune cells and cause release of inflammatory cytokines and chemokines that can baffle the host. These inflammatory markers, if unchecked, could lead to serious health issues and even death.

It is interesting to note that these set of molecules can trigger massive T-cell proliferation like a mitogen, but its MHC class II dependency remains a pre-requisite. Unlike the conventional antigens, superantigens (SAg) do not require pre-processing and display onto Antigen presenting cells (APC). They form MHC class-II/SAg/TCR complex by binding to the Variable β (Vβ) subunit of T-cells and thus lead to non-specific proliferation of sub set of T-cells. Furthermore, it is seen that SAgs bring different sub-types of cells closer, thus compelling them to exchange signals that lead to biochemical changes and thus triggering inflammatory cytokines release. But, it must be kept in mind that even if every superantigen has similar structure and function, they vary significantly based on host genetics and environmental factors. Studying one does not mean studying them all. The fact that *Streptococcus pyogenes* has eleven different kinds of superantigens suggests the non-redundancy in functioning of these proteins.

There have been many studies done about superantigens, thus pointing to the fact that this is indeed an advancing field that still lures a lot of scientists all over the world. Sophisticated techniques and bioinformatics have also deciphered new aspects to the study and have aided in understanding the role of these small yet potent molecules in pathogenesis of *S. pyogenes*. These new findings have paved new ways for in-depth studies in mechanisms and functionality of these proteins.

This book starts with a basic knowledge about *S. pyogenes* as a pathogen. It elucidates briefly about the array of virulence factors possessed by *S. pyogenes*. These help in evading host immune responses such as by the activation of non specific T-cell subpopulations, cleaving of IgG, degradation of chemokines and inactivation of complement system. It also releases DNases to chop down neutrophil entrapments (NETs) and proteases for cleaving antimicrobial peptides and proteins. Out of all the virulence factors, this book mainly targets superantigens and explains how they are different from conventional antigens. It is interesting to see that despite having differences in amino acids, these proteins form similar structure to carry out their functions. Moreover, this book also elaborates those diseases in which superantigens are actively involved. In addition, useful aspects of superantigens and different therapeutic interventions to eradicate superantigens induced diseased have also been discussed. Overall, this book is an attempt to provide ample knowledge and better understanding about *S. pyogenes* and their superantigens for readers. To develop easy understanding, many illustrative figures have also been used to explain the text.

Contents

Abstract

This book starts with a basic knowledge about *Streptococcus pyogenes* as a pathogen. It elucidates briefly about the array of virulence factors possessed by *S. pyogenes*. These help in evading host immune responses such as by the activation of non-specific T-cell subpopulations, cleaving of IgG, degradation of chemokines and inactivation of complement system. It also releases DNases to chop down neutrophil entrapments (NETs) and proteases for cleaving antimicrobial peptides and proteins. Out of all the virulence factors, this book mainly targets superantigens and explains how they are different from conventional antigens. These molecules can trigger massive T-cell proliferation like a mitogen, but its MHC class II dependency remains a pre-requisite. Unlike the conventional antigens, superantigens (SAg) do not require pre-processing and display onto Antigen presenting cells (APC). They form MHC class-II/SAg/TCR complex by binding to the Variable β (Vβ) subunit of T-cells and thus lead to non-specific proliferation of sub set of T-cells. Furthermore, it is seen that SAgs bring different sub-types of cells closer, thus compelling them to exchange signals that lead to biochemical changes and thus triggering inflammatory cytokines release. Moreover, this book elaborates those diseases in which superantigens are actively involved. In addition, useful aspects of superantigens and different therapeutic interventions to eradicate superantigens induced diseased have also been discussed in this book. Overall, this book is an attempt to provide ample knowledge and better understanding about *Streptococcus pyogenes* and their superantigens for readers. To develop easy understanding, many illustrative figures have also been used to explain the text.

Abstract

Streptococcal Superantigens

Abstract This book starts with a basic knowledge about *Streptococcus pyogenes* as a pathogen. It elucidates briefly about the array of virulence factors possessed by *Streptococcus pyogenes*. These help in evading host immune responses such as by the activation of nonspecific T-cell subpopulations, cleaving of IgG, degradation of chemokines and inactivation of complement system. It also releases DNases to chop down neutrophil entrapments (NETs) and proteases for cleaving antimicrobial peptides and proteins. Out of all the virulence factors, this book mainly targets superantigens and explains how they are different from conventional antigens. These molecules can trigger massive T-cell proliferation like a mitogen, but its MHC class II dependency remains a prerequisite. Unlike the conventional antigens, superantigens (SAg) do not require preprocessing and display onto Antigen pre- senting cells (APC). They form MHC class II/SAg/TCR complex by binding to the variable β (Vβ) subunit of T-cells and thus lead to nonspecific proliferation of subset of T-cells. Furthermore, it is seen that SAgs bring different subtypes of cells closer, thus compelling them to exchange signals that lead to biochemical changes and thus triggering inflammatory cytokines release. Moreover, this book also elaborates those diseases in which superantigens are actively involved. In addition, useful aspects of superantigens and different therapeutic interventions to eradicate superantigens induced diseased have also been discussed in this book. Overall, this book is an attempt to provide ample knowledge and better understanding about *Streptococcus pyogenes* and their superantigens for readers. To develop easy understanding, many illustrative figures have also been used to explain the text.

Keywords *Streptococcus pyogenes* · Superantigens · T-cell · MHC class II · Toxic shock syndrome · Intravenous immunoglobulins (IVIG)

© The Author(s) 2015
A. Babbar, *Streptococcal Superantigens*,
SpringerBriefs in Microbiology, DOI 10.1007/978-3-319-22455-8_1

1   *Streptococcus pyogenes*

1.1   *Introduction*

The genus *Streptococcus* includes Gram-positive bacteria that belong to the phylum Firmicutes. Since the cell division takes place along one axis, they generally grow in chains, hence the name '*Streptococcus*' from the Greek word 'Streptos' which mean 'easily twisted or bend'. All *Streptococci* are generally negative for catalase and oxidase and thus are facultative anaerobes. *Streptococcus* can be differentiated into different species by conventional biochemical testing or modern 16s rRNA sequencing methods. These species can furthermore be classified based on their hemolysis (alpha, beta or gamma) and surface antigens (A, B, C, D, F, G, and H). Surface antigen identification is a well-characterized method used in typing of Streptococcus. This method was developed by Rebecca Lancefield and hence the name of this method is Lancefield typing. It identifies the carbohydrate linkage on the surface of the bacteria that composes of *N*-acetyl glucosamine linked to a rhamnose polymer backbone.

Beta-hemolytic Group A streptococcus (GAS) or *Streptococcus pyogenes* is an important human pathogen causing wide spectrum of diseases ranging from mild infections of skin and mucosa to severe invasive infections generally associated with high mortality, such as necrotizing soft tissue infections and toxic shock syndrome (Cunningham 2000). Repeated exposure can lead to serious post-infection immune sequelae, including rheumatic fever and acute glomerulo-nephritis. The global burden of *S. pyogenes* is very high. Worldwide, 800 million infections are reported, resulting in about 1 million deaths every year (Carapetis et al. 2005). Although *S. pyogenes* is sensitive to penicillin, yet there have been outbreaks observed in the past in many regions of the world (Lamagni et al. 2005; Smoot 2002). There have been different methods for typing of *S. pyogenes* reported in the literature. The most widely used serological method is the *emm* typing (deciphering the M-protein subtype). M-protein is an important surface-bound protein of *S. pyogenes* and is reported to be involved in pathogenesis (Bisno et al. 2003). It can be conventionally tested either by extracting the M-protein after boiling the bacteria in concentrated hydrochloric acid or through amplification and sequencing of the *emm* gene responsible for production of M-protein. In the year 1946, isolates of *S. pyogenes* were classified based on their surface T antigen (Lancefield and Dole 1946). Till date, around 220 different serotypes of M-proteins and 20 different T-Types have been reported. Assigning of a sequence to a particular serotype of M-protein is done as per the standard protocol of Center for Disease Control and Prevention (CDC) (http://www.cdc.gov/streplab/M-ProteinGene-typing.html).

1.2 Epidemiology of Streptococcal Infections

Streptococcus pyogenes is a common cause of a wide spectrum of infections in human; being not limited to a specific age group. It causes infections in infants, children and adults. Infections of *S. pyogenes* date back to 5th century B.C (LeFebvre 2008). *Streptococcus pyogenes* infections have long been associated with significant morbidity and mortality, but toward 1950s, a noticeable decline in the incidences and severity of such infections was observed. However, over the last years, there has been marked increase in the incidence of severe invasive *S. pyogenes* infections (Steven 1992). These include necrotizing fasciitis, toxic shock syndrome, and streptococcal bacteremia. As per a report from the Centers for Disease Control and Prevention (CDC), approximately 9,000–11,500 cases of invasive GAS infections occur every year in the United States; from which, Streptococcal Toxic shock syndrome and necrotizing fasciitis each accounted for approximately 6–7 % of cases. More than 10 million noninvasive GAS infections (primarily throat and superficial skin infections) occur annually (Deutscher et al. 2011). The renaissance of GAS as 'flesh eating bacteria' was well documented in the United States and Europe in the 1980s and into the 1990s, gaining a lot of public awareness. Continuous resurrection, increased mortality and a lack of therapeutic vaccine against *S. pyogenes* remains a major concern. In Denmark, number of cases reported for rheumatic fever decreased from 250 (in 1862) to 100 cases (in 1962) per 100,000 individuals. By 1980, the incidence further decreased to 0.23–1.88 cases per 100,000 individuals. The incidence of post-streptococcal glomerulonephritis (PSGN) ranges from 10–28.5 new cases per 100,000 individuals per year. PSGN sums to 2.6–3.7 % of all glomerulopathies within the period of 1987–1992, but only 9 new cases were reported from 1992 to 1994. In China and Singapore too, the incidence of PSGN has declined within the past 40 years. In Mexico, the combined data from 2 hospitals showed a reduction in cases of PSGN from 27 in 1992 to only 6 in 2003 (Rodriguez-Iturbe and Musser 2008). An initiative called 'Strep-EURO' program analyzed number of cases reported from 11 participating countries, thus investigated the epidemiology of severe *S. pyogenes* disease in Europe during the 2000s. Approximately, 2.46 cases per 100,000 individuals was reported in Finland; 2.58 per 100,000 individuals, in Denmark; 3.1 per 100,000 individuals, in Sweden; and 3.31 per 100,000 individuals in the United Kingdom (Eurosurveillance 2005). In contrast, the incidences reported in southern Europe—the Czech Republic, Romania, and Italy—were substantially lower (0.3–1.5 per 100,000 individuals). It was associated with poor microbiologic investigative methods in these countries. There is also a lack of epidemiological data in many of the developing countries; including India. This is an issue of great concern as *S. pyogenes* is epidemic. A study showed that the most common clinical presentations were bacteremia without focus (30 %), pneumonia (28 %), and cellulitis (17 %) (Haggar et al. 2012).

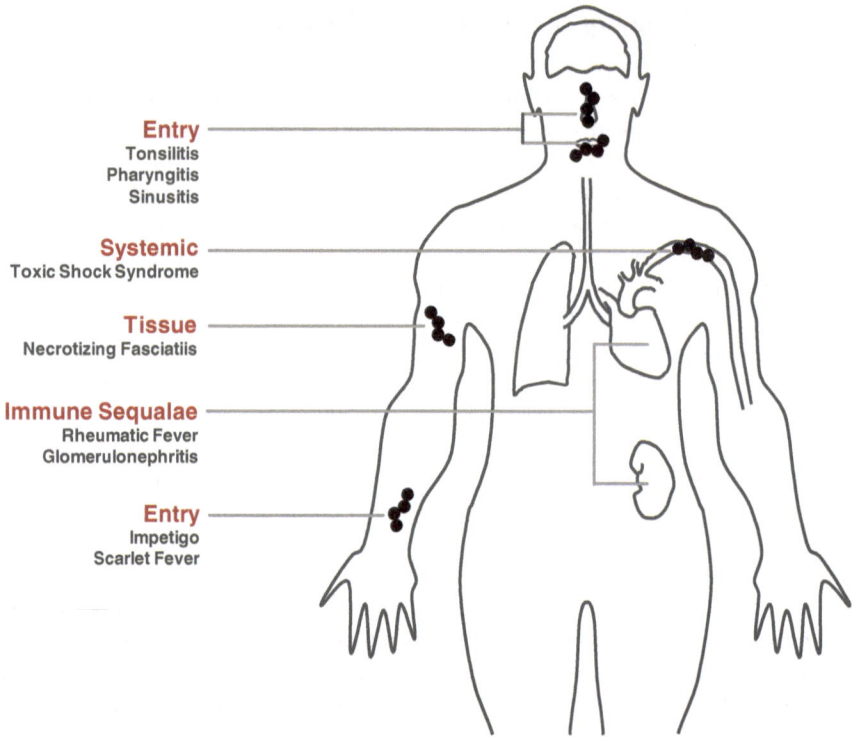

Entry
Tonsilitis
Pharyngitis
Sinusitis

Systemic
Toxic Shock Syndrome

Tissue
Necrotizing Fasciatiis

Immune Sequalae
Rheumatic Fever
Glomerulonephritis

Entry
Impetigo
Scarlet Fever

Fig. 1 Diseases caused by *S. pyogenes*

1.3   Clinical Manifestations

Streptococcus pyogenes is a leading cause for a wide spectrum of diseases. It can cause superficial to invasive skin infections. It depends not only on the niche it survives in the body but also the serotype of the bacteria. Some of the serotypes are predominant in causing only a specific kind of disease. Figure 1 classifies the different aspects of the diseases caused by *S. pyogenes*. Out of the wide range of clinical manifestations caused by this pathogen, some of the features of the sup-purative infections are described here.

1.3.1   Pharyngitis and Scarlet Fever

Pharyngitis (or commonly known as Strep throat) is the most common infection among children (Shaikh et al. 2010). It targets mainly the pharynx including tonsils. General symptoms include fever, sore throat, and tonsillitis. Since it is a contagious disease, it is not restricted to the age group of children but can spread to adults too. Pharyngitis can lead to post-infection immune sequelae such as rheumatic fever.

Both of these diseases are seasonal and the cases peak particularly during autumn and winter (Pichichero and Casey 2007). Certain M-protein serotypes, such as M-types 1, 3, 5, 6, 14, 18, 19, and 24 of *S. pyogenes*, are found associated with pharyngitis and rheumatic fever. Scarlet fever, on the other hand, shows similar symptoms in children like sore throat, fever, and a characteristic red rash. There is no vaccine developed for scarlet fever so far, but it can be efficiently treated by antibiotics. The scars produced are mainly due to erythrogenic toxin (SpeA, SpeB, and SpeC). SpeA and SpeC are bacteriophagal, while SpeB is chromosomally encoded. SpeA and SpeC are also known as superantigens (described later in much detail), are responsible for rash, strawberry tongue, and desquamation of skin seen in scarlet fever.

1.3.2 Pyoderma and Skin Infections

Pyoderma refers to any infection that is pyrogenic, being associated with pus formation. It affects more than 111 million children worldwide, which makes it one of the three most common skin disorders in children (Andrews et al. 2009). It includes superficial infections like impetigo. Impetigo is a highly contagious bacterial infection that generally targets the school children. Different serotypes (M-types) are involved in pyoderma and skin infections than throat infection (Cunningham 2000). The patient generally develops red sores on the face and neck. It may clear on its own in 2–3 weeks but antibiotics can shorten the course of disease.

Group A streptococcal strains may enter the skin through abrasions and other types of lesions to penetrate the epidermis and produce erysipelas or cellulitis. Erysipelas is a distinctive form of cellulitis with characteristically raised and erythematous superficial layers of the skin, while cellulitis affects subcutaneous tissues (Bonnetblanc and Bédane 2003). Cellulitis may occur from infected burns or wounds. Both erysipelas and cellulitis can be caused by streptococcal groups A, B, C, and G.

1.3.3 Invasive Diseases: Toxic Shock Syndrome and Necrotizing Fasciitis

Streptococcus pyogenes sometimes causes diseases that go deep into the tissue or becomes systemic. Some of the invasive diseases caused by *S. pyogenes* include Streptococcal Toxic Shock Syndrome (STSS) or rapidly progressing Necrotizing Soft Tissue Infections (NSTIs) like necrotizing fasciitis. Also known as flesh-eating disease, flesh-eating bacteria or flesh-eating bacteria syndrome (Rapini et al. 2007), it is a rare infection of the deeper layers of skin and subcutaneous tissue, spreading across the fascia within the subcutaneous tissue. The most common and persistent symptom of necrotizing fasciitis is necrosis of the subcutaneous tissue and fascia. It was first described in 1952, as necrosis with relative sparing of the underlying muscle (Wilson 1952). NSTIs are generally associated with medical risk factors like diabetes mellitus, cardiovascular disease, or recent surgery, but also with

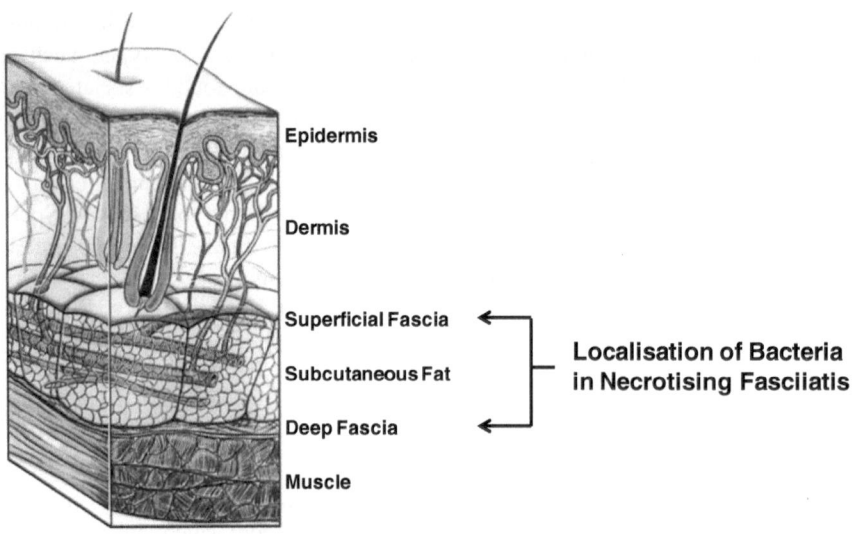

Epidermis

Dermis

Superficial Fascia

Subcutaneous Fat

Deep Fascia

Muscle

Localisation of Bacteria in Necrotising Fasciiatis

Fig. 2 Necrotizing Fasciitis

small wound infections or trauma and have a mortality rate of 12–35 %. NSTI-affected patients need intensive care unit treatment, rigorous medication, surgery to remove infected tissue, and specialized treatment strategies like hyperbaric oxygen therapy (HBO) and intravenous immunoglobulin administration (IVIG). IVIG is a sterile preparation of concentrated antibodies recovered from the pooled serum of at least 1000 healthy donors that is discussed to work by a variety of mechanisms like activation of complement, facilitation of opsonization, and neutralization of bacterial toxins. It is known that *S. pyogenes* secrete highly potent exotoxins with superantigen activity causing systemic toxicity and hyperinflammatory response (Fig. 2) and neutralising antibodies in IVIG are thought to block their activity.

On the other hand, Streptococcal toxic shock syndrome was characterized by hypotension and multiple organ failure. It is due to the exotoxins (basically superantigens) released by the bacteria that lead to hyper activation of T-cells. Admission in intensive care unit is often required, especially in case of multiple organ failure.

1.3.4   Post-Infection Immune Sequelae: Rheumatic Fever and Glomerulonephritis (PSGN)

Rheumatic fever is an inflammatory disease that may develop after an infection with *S. pyogenes* bacteria (such as strep throat or scarlet fever). The disease can affect the heart, joints, skin, and brain. The rheumatic heart disease can be diagnosed for the evidence of Streptococcal infection through an increase in anti-streptolysin O or

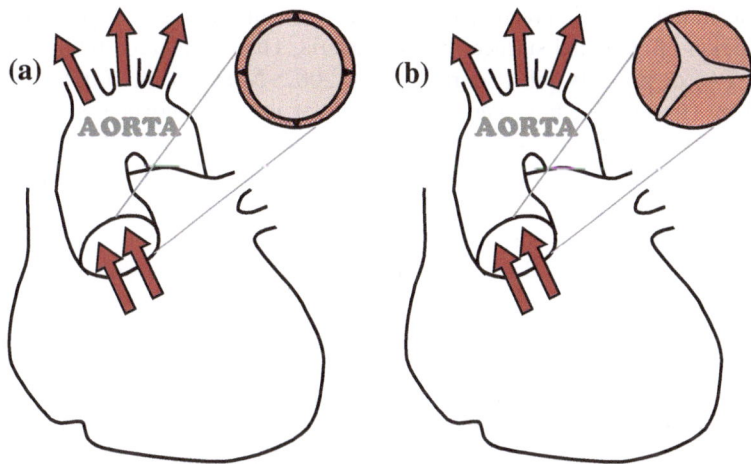

Fig. 3 a Normal heart. **b** Rheumatic heart

anti-DNases antibodies found in the serum of the patient (Liu et al. 2015; Jones criteria 1992). The echocardiographic and Doppler (E&D) studies have identified subclinical carditis in patients with acute rheumatic fever which suggested RHD in heart patients who had isolated cases of sydenham chorea (Kumar and Tandon 2013). Major symptoms include polyarthritis, carditis inflammation of the heart muscle (Myocarditis), pericarditis, subcutaneous nodules painless, and firm collections of collagen fibers over bones or tendons, and erythema marginatum. The minor criteria are Rheumatic fever of 38.2–38.9 °C (100.8–102.0 °F), arthralgia joint pain without swelling (only if polyarthritis is not present as a major symptom), raised erythrocyte sedimentation rate or C-reactive protein, leukocytosis, ECG showing features of heart block, such as a prolonged PR interval (Fig. 3).

Similarly, post-streptococcal glomerulonephritis (PSGN) is also one of the post-infection sequelae of *S. pyogenes*. Symptoms of PSGN vary from asymptomatic, microscopic hematuria to the full-blown acute nephritic syndrome, characterized by red to brown urine, proteinuria, edema, hypertension, and acute kidney injury. The prognosis is generally favorable, especially in children, but in some cases, the long-term prognosis is not benign.

1.4 Virulence Factors of Streptococcus pyogenes

Invasive infections of *S. pyogenes* often require colonization of bacteria (as in case of NSTIs) or their dissemination (systemic infections like TSS). For proper invasion and colonization, bacteria have to face the immune system of the host. In order to defend and protect themselves, the bacteria release lot of factors to combat the immune system and these factors helps in proliferation and dissemination of the bacteria. Initial weak interaction is initiated with cell surface or mucosa through colonization of epithelial surfaces by the bacteria. This is triggered by lipoteichoic acid (LTA) or possibly pili (Courtney et al. 2002; Nobbs et al. 2009). It is then pursed by stronger binding (carbohydrate and/or protein) between epithelial cell layer and the bacteria. A lot of proteins like M-protein, SfbI, SfbII, and ECM-binding proteins seem to be involved in this process. *Streptococcus pyogenes* has developed a mechanism to counter neutrophil extracellular traps (NETs), invading phagocytosis, nonspecific T-cell proliferation, cleavage of IL-8, degradation of IgG, and so on. Some of the factors that have roles at different stages of infections are mentioned below:

(1) Hyaluronic Acid Capsule

Hyaluronic acid is naturally present in human body. Interestingly, many *S. pyogenes* strains are surrounded by hyaluronic acid capsule (Chong et al. 2005). This capsule helps the bacteria to invade phagocytosis (Cunningham 2000); thus aiding the bacteria to survive in the host. It is also seen as an important factor in adhering as it binds to CD44 receptors on the epithelial cells (Schrager et al. 1998). Hyaluronic acid is a linear polymer of glucuronic-β-1, 3-*N*-acetylglucosamine, produced by highly conserved has *ABC* (hyaluron synthase) operon. It is also seen to be important in survival of bacteria in extracellular traps made by neutrophils (NETs) by reticence of LL-37 (antimicrobial peptide that is an important contributor in formation of NETs) (Cole et al. 2010) (Fig. 4).

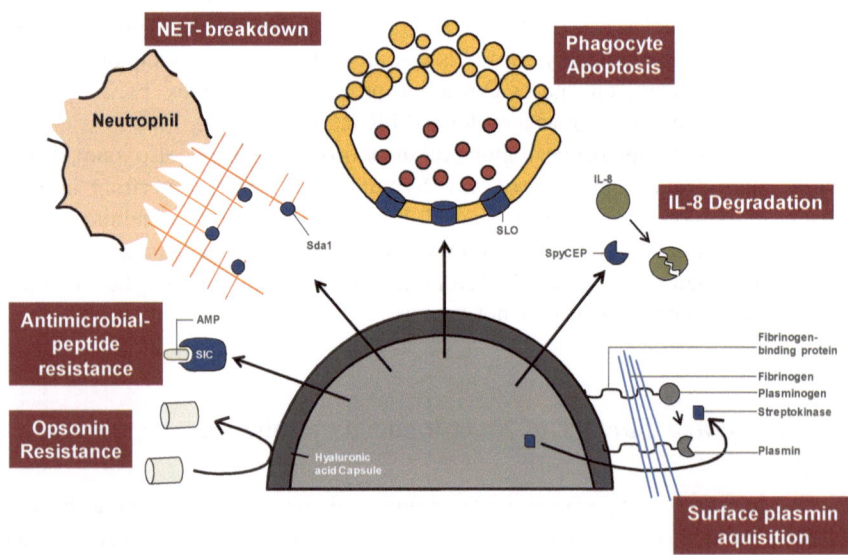

Fig. 4 Virulence factors of *S. pyogenes* for disabling Neutrophils [Modified from Cole et al. (2011)]

(2) M-protein

M-protein was first described by Lancefield in 1928. It is a surface-bound protein and the primary antigen of *S. pyogenes* (Metzgar and Zampolli 2011). It is an important factor in initial attachment of the bacterial cells (Frick et al. 2003). It also contributes to the antiphagocytic function by binding to fibrinogen (Herwald et al. 2004). The M-protein is hypervariable at its N-terminal and has been basis for serological differentiation of *S. pyogenes*. Different M-types have been seen associated with different clinical manifestations. Consequently, different serotypes behave divergently when injected into various mouse models. Also, immunization with M-protein provides a strong protectivity towards streptococcal infections (Steer et al. 2009; Dale and Chiang 1995). M-protein, hence, are the most important candidate for vaccine development. A 30 valent vaccine, that covers all the M-types, is seen to generate cross-opsonic antibodies against nonvaccine serotypes of *S. pyogenes* (Dale et al. 2011).M-protein

(3) Streptolysin—O (SLO)

69 kDa streptolysin O (SLO) is a thiol-activated, cholesterol-dependent cytolysin that forms large pores of about 30 nm in the host cell membrane leading to apoptosis of the phagocyte (Nilsson et al. 2006; Bryant et al. 2005). It is reported to cause apoptosis in epithelial cells, neutrophils, and macrophages (Timmer et al. 2009). SLO also aids infiltration of superantigens into stratified mucosa (Brosnahan et al. 2009). Since SLO is co-transcribed with NAD glycohydrolase, it is thus seen to mediate the translocation of NAD glycohydrolase into cells through the pore being formed. NAD glycohydrolase modulates the host cell signaling pathways; hence assisting in *S. pyogenes* internalization (Madden et al. 2001). SLO expression is noticed to be higher in severe invasive cases when compared to noninvasive controls (Ato et al. 2008).M-protein

(4) SpyCEP (IL-8 Protease)

SpyCEP (also known as ScpC, Spy0416, CepA, Prts, or SP24) is one of the virulence factors that are upregulated in invasive diseases (Sumby et al. 2006). SpyCEP can cleave various chemokines like Interleukin-8 (IL-8), Grap-2 (GRB2-related adapter protein 2), and GCP-2 (granulocyte chemotactic peptide-2) (Sumby et al. 2006; Chiappini et al. 2012). It is a subtilisin-like protease being secreted by *S. pyogenes* (Edwards et al. 2005). A mature SpyCEP is a dimer made from 2 subunits of 30 and 150 kDa. A stable interaction between both the subunits is a prerequisite for enzymatic activity of the protease (Zingaretti et al. 2010). Recently, SpyCEP is seen to be inimical for biofilm formation of the bacteria (Andreoni et al. 2014).M-protein

(5) Streptococcal inhibitor of complement (SIC)

SIC is a highly polymorphic 31-kDa protein being secreted out of *S. pyogenes* that inferences with formation of Membrane Attack Complex (MAC) by constraining the interaction of complement complex C5b67 with host cell membrane

(Fernie-King et al. 2001). It is also reported to inhibit other immune system molecules like human cathelicidin LL-37, α-defensins, secretory leukocyte protease inhibitor, and lysozyme (Cole et al. 2011).

(6) IgG endopeptidase (Mac/IdeS)

IgG endopeptidase (IdeS; also known as Mac, MspA) is a cysteine protease that specifically cleaves human immunoglobulin (IgG) (Agniswamy et al. 2004). It is a homologue of human leukocyte ß2 integrins which are reported to inhibit the activation of neutrophils and production of reactive oxygen species by binding CD16B (a low-affinity Fc receptor) (Cole et al. 2011). Two different variants of Mac have been described. Complex-I and Complex-II differ in amino acid level (amino acids 112 to 205) (Lei et al. 2002). Complex-I wields their function through proteolytic cleavage of IgG (Agniswamy et al. 2004); whereas Complex-II, which is a weaker endopeptidase, blocks interaction of FcγIIIb receptor; thus blocking phagocytic killing (Agniswamy et al. 2004; Lei et al. 2002).

(7) Streptokinase (Ska)

A common host defense mechanism is to trap the bacteria within the fibrin clot in order to block the dissemination of the bacteria. *Streptococcus pyogenes* can very cleverly disable this host defense mechanism by production of Streptokinase (Ska). Ska activates plasminogen, a 92-kDa glycoprotein found in plasma that is an important factor in fibrinolysis (Dano et al. 1985). Plasminogen is converted to plasmin that dissolves the fibrin clot thus letting the bacteria 'out' or spreading (Ponting et al. 1992). Ska polymorphism is well documented. There are three different clusters of Ska–Cluster 1, Cluster 2a, and Cluster 2b—that have different modes of actions for plasminogen activation (McArthur et al. 2008). All strains of *S. pyogenes* are known to secrete streptokinase and are capable for acquisition of cell surface plasmin (McArthur et al. 2008).

(8) Extracellular streptodornase D (sda1)

In order to promote the entrapment and consequent clearance of bacteria, neutrophils release 'NETs' (Neutrophil Extracellular Traps) that are made from DNA and proteins (Brinkmann et al. 2004). Interesting, *S. pyogenes* releases an extracellular DNase called Sda1 that chops off the DNA network of NETs, thus helping the bacteria from the entrapment and killing (Buchanan et al. 2006). Recently, it has been reported that Sda1 helps the bacterial evasion through by TLR9-dependent recognition (Uchiyama et al. 2012). It is done through autodegradation of CpG-rich DNA by bacterial DNase (Uchiyama et al. 2012).

(9) SpeB

SpeB is a broad-spectrum cysteine protease released by most of the *S. pyogenes* strains (Haataja and Gerlach 2001). Although almost all (<99 %) of the strains have this gene, there is a considerable difference seen in its expression levels (Chaussee et al. 1996). SpeB is initially produced as a 40-kDa zymogen which is then self-cleaved to generate a 23-kDa mature form (Carroll and Musser 2011). Loss of

SpeB is a crucial factor for conversion of a 'local infection' causing strain to be invasive. As SpeB is a broad-spectrum protease, it chops off its own virulence factors being released. Once there is a mutation in covRS operon, activity of SpeB goes down. This enables the bacteria to produce more functionally active exotoxins, hence increasing the virulence of the strain (Turner and Sriskandan 2007).

(10) Superantigens

Streptococcus pyogenes secrete number of Streptococcal pyogenic exotoxins (SPEs), that share sequence similarity of about 17–48 %, nevertheless are structurally related (Cunningham 2000; Kotb 1995). These exotoxins belong to the family of 'Superantigens' (SAgs) (Alouf et al. 1999). SAgs are small, but highly mitogenic protein molecules that bind to Major Histocompatibility Complex (MHC) class II antigens and TcR molecules leading to co-stimulation of large number of T-cells (Proft et al. 2000). They are known to bind MHC class II outside the peptide groove and to CDR2 and HV4 regions on TcR Vß chains (Sundberg et al. 2002). These superantigens are seen to surpass the conventional processing by immune cells and hence cause massive cytokine release (Proft and Fraser 2007). Several SAg genes are located on mobile DNA elements, such as bacteriophages, integrated in bacterial genomes (Ferretti et al. 2001) and hence, are more likely to be genetically transferred among different streptococcal species (Kalia and Bessen 2003). Superantigens have been discussed in detail in the following section.

2 Superantigens (SAg)

2.1 Introduction

In recent years, tremendous attention is being given to group of proteins known as 'superantigens'. The way these molecules communicate with the immune system has been seemingly captivating. Unlike the conventional antigens, superantigens (SAg) are small, but highly mitogenic proteins that specifically bind to Major Histocompatibility Complex (MHC) class II antigens and TCR molecules triggering enormous co-stimulation of T-cells (Fraser et al. 2000). Possessing superantigens are not restricted only to a certain subset of pathogens but are produced by viruses and bacteria including *Streptococcus*, *Staphylococcus*, *Plasmodium falciparum* and *Mycoplasma* species. Some of the superantigens are membrane bound while others are secreted as exotoxins. These molecules interact with the host immune system in a nonconventional way and potentially trigger immune sequelae such as toxic shock syndrome and autoimmunity. Understanding the role of superantigens in disease induction has helped in deciphering the pathogenesis of bacteria producing it and has initiated exploration of new efficient therapeutic strategies. This chapter focuses mainly on bacterial pyrogenic superantigens secreted by *S. pyogenes* (Fig. 5) and their role in bacterial pathogenesis and disease manifestation.

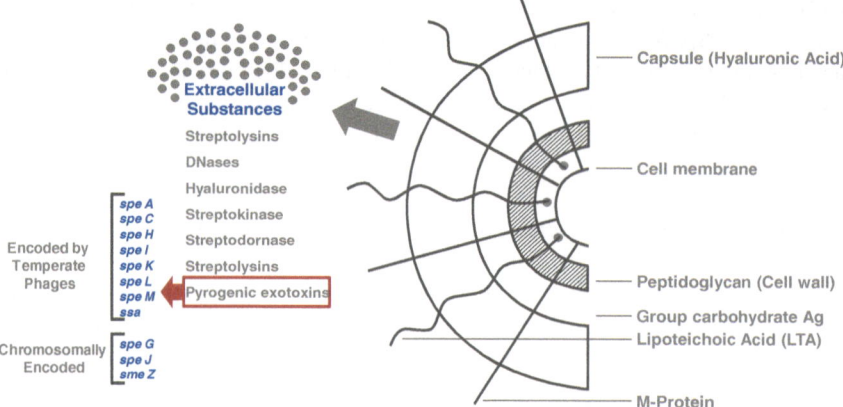

Fig. 5 Superantigens released by *S. pyogenes*

2.2 Detection of Superantigens

Identification of superantigens can be demarcated in two distinct clades: the pre-genomic era and the post genomic era. The pre-genomic era constituted of identification of superantigens by their functional assays whereas the post genomic era identified these proteins by scanning the whole genome sequences available for *S. pyogenes*. The *S. pyogenes* superantigens can be found both on chromosome as well as on integrated bacteriophage regions. Out of 11 superantigens known for *S. pyogenes*, 3 (SmeZ, SpeG, and SpeJ) are chromosomally encoded. Rest is associated with temperate phages, which is the mechanism for their horizontal transfer. Identification and characterization of all superantigens in these two clades has been explained below.

2.2.1 Pre-genomics Era

Discovery of streptococcal superantigens dates back to the year 1924 when culture filtrates of *S. pyogenes* injected subcutaneously caused 'scarlatina' rash. This reaction was named as 'Dick's reaction' (Dick and Dick 1924). It was speculated to be caused by erythrogenic toxin that was later named as "scarlet fever toxin" or Streptococcal pyogenic toxin A (**SpeA**). Streptococcal pyogenic toxin B (**SpeB**) was identified in 1934 by Hooker and Follensby and was characterized by Stock and Lynn in 1969. Later, SpeB was cloned in M12 GAS strain and was found to be functionally similar to streptococcal cysteine protease (SCP) (Bohach et al. 1988; Gerlach et al. 1983). Third toxin (Streptococcal pyogenic exotoxin C; **SpeC**) was isolated and characterized in year 1960 from culture filtrate of scarlet fever causing strain (Cremer and Watson 1960). All three toxins caused massive expansion of T-cells and were pyrogenic in rabbit. Hence, these toxins were named as 'pyrogenic exotoxins'. Even though SpeB was seen inducing T-cell subpopulation, it was found to be a contamination of another potent superantigen that was later named as Streptococcal mitogenic exotoxin Z (**SmeZ**). SmeZ was first identified as potent Vβ4 and 8 simulating superantigen from M1 serotype (Kamezawa et al. 1997). Later, an allelic variant of SmeZ, known as **SmeZ-2** was found. It was purified from streptococcal cultures fostering T12 phage. The gene for SmeZ-2 was later cloned in year 1999 (Proft et al. 1999). Streptococcal superantigen (SSA) was found out in the year 1993 from M3 serotype strain of *S. pyogenes* (Mollick et al. 1993).

2.2.2 Post-genomics Era

The genomic revolution simplified the struggle to identify new superantigens. Even though streptococcal and staphylococcal SAgs share a limited DNA sequence homology of about 17–48 %, they share two strongly conserved 'Family signature motifs' (STAPH_STREP_TOXIN) as determined by PROSITE **Y-G-G-[LIV]-T-X (4)-N** (PS00277) and **K-X(2)-[LIVF]-X(4)-[LIVF]-D-X(3)-R-X(2)-L-X(5)-[LIV]-Y** (PS00278). Scanning the genome for these signature motifs led to the discovery of **SpeG, SpeH, SpeI**, and **SpeJ** (Proft et al. 1999). The SpeL gene was identified in serotype M3 on temperate phage ΦNIH1.1 (Ikebe et al. 2002). A lot of confusions with the nomenclature of SpeK, SpeL, and SpeM arose. SpeK was first described as a pseudogene with partial open reading frame in M1 serotype by Ferretti et al. (2001). This gene was later named as SpeK-M1 (Smoot et al. 2002). The name SpeK was also given to the gene designated as SpeL by Ikebe et al. (2002). Also SpeM was designated as SpeL/M by Proft and Fraser (2007). In order to avoid any confusion with the nomenclature, the names of the genes as suggested by Commons et al. (2014) are followed.

2.2.3 Diversity in Streptococcal Superantigen Genes

By analyzing the nucleotide sequences for 'superantigens' in the nucleotide database, Commons et al. could find nearly 91 different alleles of genes termed as superantigens of the streptococcal family (Commons et al. 2014). The number of nucleotide alleles varied from minimum of 3 (SpeI, SpeJ) to maximum of 53 (SmeZ). SmeZ is the most variable superantigen. The difference in amino acid sequences from the first published SmeZ (known as SmeZ-1) and SmeZ-2 is of a pentapeptide 96-EEPMS-100 to 96-KTSIP-100. This pentapeptide represents a potential B-cell epitope and hence, the allelic difference between SmeZ-1 and SmeZ-2 is the greatest difference in SmeZ alleles. Despite the allelic variations among SmeZ, the MHC II and Vβ specificity remains the same. This shows that the allelic variations emanated in order to escape the antibody selection and neutralization. Apart from SmeZ, SpeA and SpeC have been seen to have minor allelic variations. SpeA2 and SpeA3 differ from SpeA1 by a single amino acid substitution (G110S in A2 and V106I in A3). Whereas, SpeA4 is approximately 9 % divergent from other 3 alleles (Reid et al. 2001). SpeC, on the other hand, has 4 allelic variations. SpeC2 differs from SpeC1 by two silent A/G transitions whereas SpeC3 and SpeC4 differ from SpeC2 by a base pair change (Norrby-Teglund et al. 2002; Kapur et al. 1992).

2.3 Mechanism of Action

Antigen presentation and antibody activation is a complex mechanism of the host to elicit a response against any foreign body or pathogen that could harm the host. Specific receptors for antigen recognition are situated on T-cell surfaces called T-cell Receptors (TCR) and on antigen presenting cells (MHC-II). Generally, antigen is cleaved in the lysosome into smaller peptides and is then displayed by the antigen presenting cells (APC). These antigen peptides finally interact/bind with T-cell that carries a specific $\alpha\beta$ TCR having constant and variable region. The selectivity of a TCR to any conventional antigen mainly depends upon the variable regions: $V\beta$, $D\beta$, $J\beta$, $V\alpha$, and $J\alpha$ (Prlic and Jameson 2002). After binding to specific TCR, peptides induce signal transduction and biochemical changes in the T-cell. Activated T-cell then proliferates to generate definite subpopulations of T-cells with peculiar variable regions and is specific to particular peptide. Unlike the conventional antigens, superantigens can bind and stimulate enormous subpopulation of T-cells (Fig. 6) via the variable region of β chain ($V\beta$ elements) (Proft and Fraser 2007).

Binding of SAg to the T-cell receptor triggers biochemical changes in the cell that leads to nonspecific activation of a copious amount of T-cells and release of inflammatory cytokines like IL-6 and IL-8. Even though superantigens activate an enormous amount of T-cells, the response is not as strong as a mitogen. A mitogen is a molecule that induces cell proliferation via mitosis. Table 1 illustrates the basic differences among an antigen, a superantigen, and a mitogen. The percentage of responding cells is higher in case of mitogen because it is not restricted to any peculiar MHC class or co-stimulatory molecules for activating T-cell population.

2.3.1 T-Cell Fate After/on Encountering Superantigen

Binding of SAg with T-cell does not necessarily actualize to its activation and proliferation. It is furthermore dependent on two distinct but dependable signals: one is the TCR binding while other is the interaction of co-stimulatory molecules that bridges APC with T-cell (Lebedeva et al. 2005). Co-stimulatory molecules such as B7 and ICAM-1 (intracellular adhesion molecule-1) interact with their ligand like CD28, thus abetting in activation and proliferation of specific subset of T-cells (Sharpe and Freeman 2002). In dearth of these important co-stimulatory signals, the T-cells "tolerate" the induction of superantigens by a process called Anergy (Schwartz 2003). Contrarily, if the level of cytokines such as TNF-α (Tumor Necrosis Factor-α) or IFN-γ (Interferon-γ) are high enough, a rearrangement in the

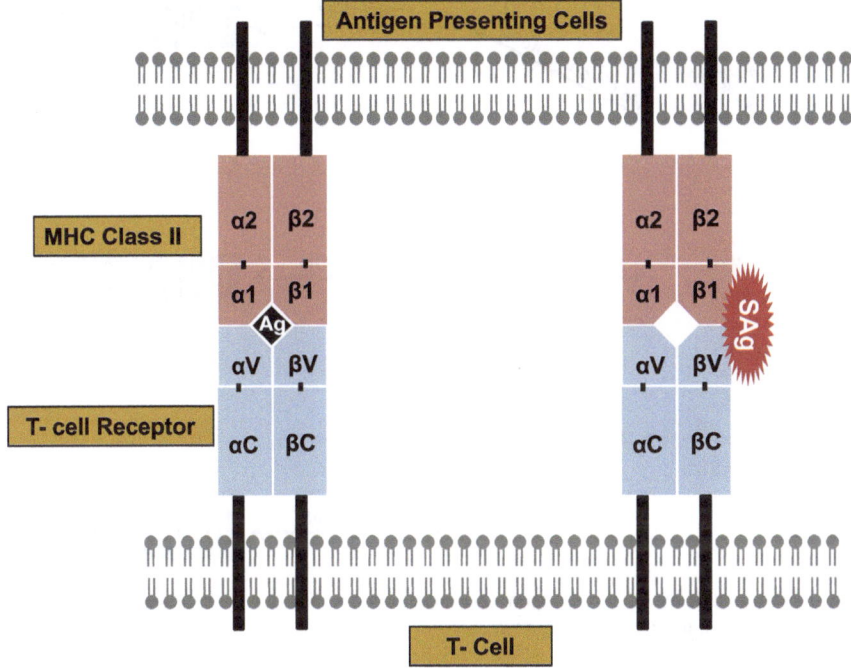

Fig. 6 Differences in binding sites of antigen (Ag) and superantigen (SAg). Ag binds to TCR via MHC-II receptors on the antigen presenting cells. Whereas, SAg bind to T-cells via variable β region, leading to nonspecific stimulation of large number of T-cells. It should also be noted that different SAgs activate different Vβ subtypes

Table 1 Differences among antigen, superantigen, and mitogen

Feature	Antigen	Superantigen	Mitogen
% of responding cells	10^{-4}–10^{-6}	5–20	80–90
Dependence on MHC class II expression	+	+	−
MHC restriction in presentation	+	−	−
Requirement for complex processing	+	−	−
Restricted Vβ usage by responding cells	−/+	+	−
Role of Vα elements	+	−/+	−
CDR3 conservation among responding cells	+	−	−

T-cell takes place. This rearrangement activates the T-cells. When a superantigen targets such activated T-cells, they program themselves to apoptosis. Hence, the effect of superantigens is eliminated (Kawabe and Ochi 1991) (Fig. 7).

Therefore, superantigen-stimulated activation and proliferation of T-cells is confined to correct binding to Vβ subunit and presence of essential co-stimulatory molecules. Activity of superantigens is dependent on specific patterns of Vβ

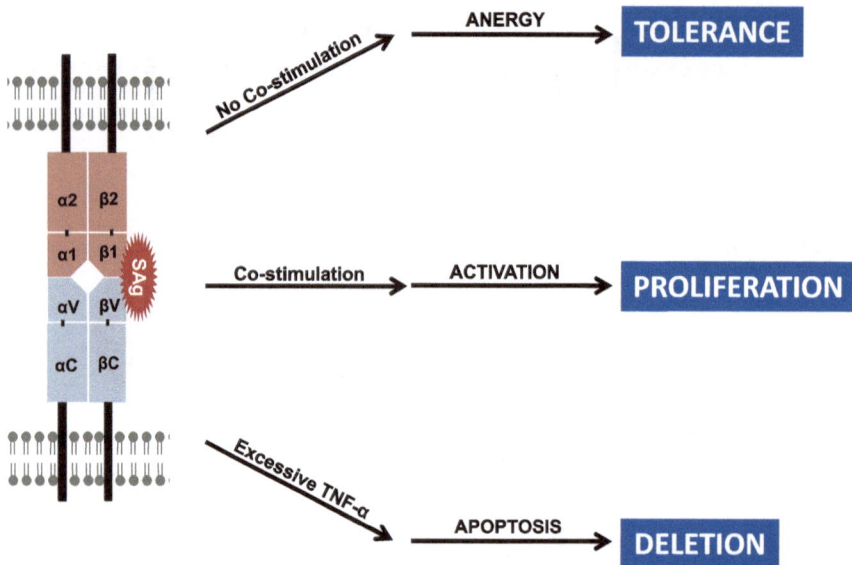

Fig. 7 Different fates of T-cells upon encounter with superantigens are decided through the interaction of various co-stimulatory molecules

regions. In case of presence of all favorable conditions, binding of superantigens causes biochemical changes in the T-cell and causes production of cytokines (like TNFα, IL-8). Some distinct changes in Vβ regions during a severe infection can hence be correlated with Vβ specificity of a particular superantigen and is considered as a clue for the involvement of the superantigen in disease manifestation and pathogenesis. The correlations and manifestations of superantigens in different diseases have been explained later in the chapter.

2.3.2   Role of CD28 in Functioning of Superantigen

As already mentioned, immune stimulation requires certain co-stimulatory molecules such as B7/CD28. C28 is one of the crucial regulators of immune system (Schwartz et al. 2001; Sharpe and Freeman 2002). CD28 is a homodimer that is constitutively expressed on the surface of T-cells. It binds to the co-stimulator B7 present on APC to initiate signaling and stimulation of T-cells (Sharpe and Freeman 2002; Riley and June 2005). B7-2 (CD86) is constitutively expressed but B7-2 (CD80) is only expressed through antigenic activation (Riley and June 2005). Therefore, activation of T-cells in early stages of infection is regulated by binding of CD28 with B7-2 (Greenwald et al. 2005; Bhatia et al. 2006). Even though it is thought just as a co-stimulatory molecule, recent data shows that superantigen can directly bind to CD28 receptor (Kaempfer et al. 2013). Blocking the CD28 receptor effectivity inhibits the toxin action.

2.3.3 Potency of Superantigen

The potency of any molecule is the minimum amount that is requisite for stimulating 50 % of T-cells of their maximum activity (P_{50}). The P_{50} value of SAg varies from as low as 0.02 pg/ml (SmeZ-2) to as high as 50 pg/ml (SpeH). SmeZ-2 has been found as the most potent with proliferation of human T-cells still seen at <0.1 fg/ml (Proft et al. 1999). *Streptococcus pyogenes* is a strict human pathogen. Still, superantigens of *S. pyogenes* can still activate T-cells from mouse and rabbit, albeit the P_{50} values were considerably low. SpeC is nonfunctional in mouse despite binding to murine I-E molecules (Li et al. 1997). There are 50 kinds of V region gene segments in the β-chain region that extend to approximately 24 families, nevertheless SAg only target a definite amount of Vβ (prevalent being Vβ2) out of all possibilities. SpeA, SpeC, SpeG, SpeH, SpeJ, and SmeZ are seen to bind to Vβ2 while SpeK, SpeL, and SpeM bind to Vβ1 (Table 2) (Proft et al. 2003a, b; Smoot et al. 2002). Interestingly, SSA and SmeZ-2 are the only SAgs that bind strongly to Vβ4 and Vβ8.

2.3.4 Biochemical Properties of Streptococcal Superantigens: Zinc Binding

To elicit a response, superantigens must concentrate enough on the antigen presenting cells. To ensure a response, superantigens have evolved to bind to MHC-II

Table 2 Functional properties of streptococcal superantigens [Modified from Proft and Fraser (2007)]

SAg	MW (kDa)	Crystal Structure	Zinc binding	MHC II binding (α/β specificity)	Human TCR Vβ specificity[a]	P_{50}^{b} (pg/ml)
Spe-A	26.0	+	+	+/−	2.1, 12.2, 14.1, 15.1	?
Spe-C	24.4	+	+	−/+	2.1, 3.2, 12.5, 15.1	0.1
Spe-G	24.6	−	+	−/+	2.1, 4.1, 6.9, 9.1, 12.3	2
Spe-H	23.6	+	+	−/+	2.1, 7.3, 9.1, 23.1	50
Spe-I	26.0	−	+	−/+	6.9, 9.1, 18.1, 22	0.1
Spe-J	24.6	+	+	−/+	2.1	0.1
Spe-K	27.4	−	+	−/+	1.1, 5.1, 23.1	1
Spe-L	26.2	−	+	−/+	1.1, 5.1, 23.1	10
Spe-M	25.3	−	+	?	1.1, 5.1, 23.1	?
SSA	26.9	−	−	+/−	1.1, 3, 15	?
SmeZ-1	24.3	−	+	−/+	2.1, 4.1, 7.3, 8.1	0.08
SmeZ 2	24.1	+	+	−/+	4.1, 8.1	0.02

[a]The major T-cell receptor Vβ targets are underlined
[b]Concentration needed for half maximum proliferation of T-cells

in mainly three distinct types: binding to MHC-II β chain, MHC-II α chain binding and dual αβ chain binding. All SAgs bind (except SpeA and SSA) bind to β chain through a coordinated Zinc atom (Zn^{2+}) at center of the interphase (Baker et al. 2001). Binding to MHC-II by SpeC was completely abrogated by addition of chelators like EDTA (Tripp et al. 2003). The co-crystallization of SpeC with HLA-DR2 possessing a peptide extracted from myelin basic protein (MBP) substantiated the zinc complex but also revealed new interactions of SpeC with the bound peptide (Li et al. 2001). It showed that affinity of Superantigen is dependent on the toxin level, MHC haplotype and sero conversion. Zinc binding motifs were also anticipated in other superantigens (SpeG, SpeH, SpeI, SpeJ, SpeK, SpeL, SpeM and SmeZ) by screening their amino acid sequences. Binding to MHC II was inhibited in all of these superantigens by addition of chelators like EDTA. All these SAgs were deficient in a universal MHC-II α-chain binding motif, a hydrophobic loop in N-terminal and thus did not fight for the binding site with SpeA and SSA (Proft et al. 1999).

2.4   Structure and Function

Superantigens are small globular proteins with a similarity of 17–48 % at nucleotide level (McCormick et al. 2001). Despite activating a range of Vβ receptors on TCR and varying nucleotide sequences, all superantigens display a conserved structural motif with two distinct domains: a smaller N-terminal domain containing two β sheets (called the OB fold) and a larger C-terminal domain having a central α-helix and a five-stranded β sheet (Baker and Acharya 2004; Papageorgiou and Acharya 2000). These domains are linked through an α helix situated on the N-terminal of protein. This structure facilitated the superantigens to bind to MHC-II and TCR at the same time as unsuccessful binding to any one of them would impair the activation of T-cells. Cross-linking of SAg with MHC-II helps in formation multivalent TCR complex by MHC-bound-SAg, thus commencing signaling.

2.4.1   Conserved Scaffold of Superantigen

Generally, SAg is made of an N-terminal helix that lades between two structural domains. Soon after the N-terminal helix is a domain (N-terminal domain) that is composed of five mixed β-barrel with Greek key topology (Zhang and Kim 2000) and is known as the OB (Oligonucleotide Oligosaccharide Binding) fold (Murzin 1993). This is followed by the C-terminal β-grasp domain through a large loop (Fig. 8). This β-grasp domain in the C-terminal is a five-stranded β-sheet that is followed by a protracted α helix. Like most of the protein, superantigens too have a hydrophobic core at its center, while other smaller hydrophobic centers are made from N-terminal helix along with the β-grasp domain. Thus making the superantigen similar to rest of the "classic" proteins: with a hydrophobic interior and

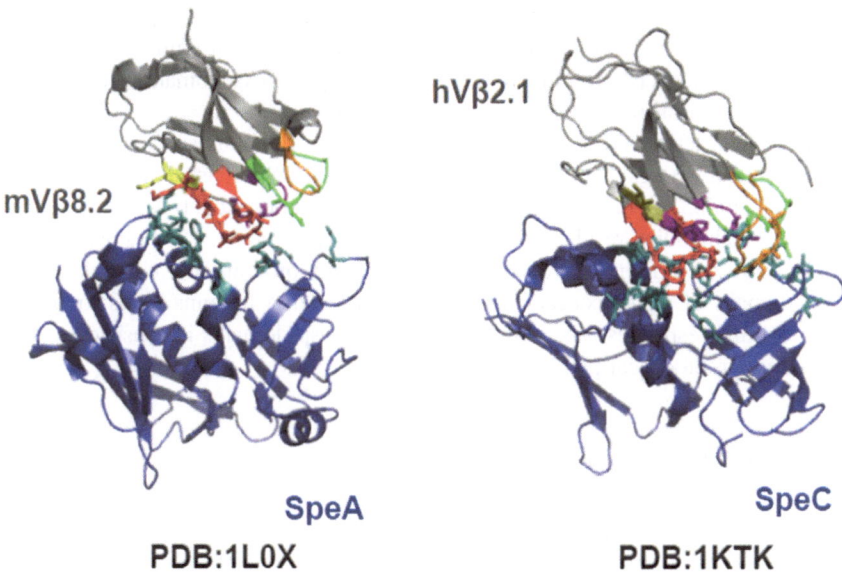

Fig. 8 Co-crystallization of SpeA (*left*) and SpeC (*right*) with their respective Vβ chains. The canonical structure of SAg consists of two domains. The N-terminal domain (*red*) consists of a β barrel motif and C-terminal domain (*blue*) consists of a β-grasp motif and an α-helix which spans the center of the structure (Sharma et al. 2014)

hydrophilic exterior. It is interesting to note that the β-grasp domain is the most eminently conserved sequences of superantigen, which defines its protein family. Many of the conserved residues are actually hydrophobic but are hidden inside the hydrophobic cores due to its folding. Interestingly, it is due to the evolutionary restrictions that these hydrophilic residues forming polar bonds are hidden in hydrophobic environment. This is because of the huge free energy cost that is paid when one residue is mutated during evolution. In case one of the hydrophilic residues is mutated, it leaves an unpaired charge that can cause free energy penalty, which can be costly for the structure as a whole. But if a hydrophobic residue is mutated, the mutation can generally be tolerated due to its low energetic cost. Hence, most of the residues that are essential or highly conserved within the superantigens are hidden inside the hydrophobic core and their conservation indicates more of its structural importance rather than functional.

The N-terminal domain of superantigen forms a β-barrel with an OB-fold structure. This OB fold is most prevalent structure and is found among different species of prokaryotes (including archaea, bacteria) and eukaryotes (Qian et al. 2001). This fold was seen to facilitate binding of different ligands on the same side of the protein (Murzin 1993). Initially OB was described in only four proteins but later on was found in more than 90 proteins and is submitted to Protein Data Bank (PDB). The SAgs use this OB loop for protein–protein interactions. SAgs like SpeA use the OB fold to bind to MHC-II α chain whereas SpeC uses this fold to oligomerise and form dimers (Roussel et al. 1997). The OB fold is quite condoning for mutations as there are just two short sequence motifs that define the OB fold. These short sequences are hidden deep inside the two SAg domains and are protected for structural rather than functional aspects.

The SAg C-terminal β-grasp domain also displays a binding surface at its mixed β-sheets. Many SAgs like SpeC, SmeZ2, SpeG and SpeJ use it for binding to MHC-II β chain. The binding, as already stated, is zinc-dependent. A Zn^{2+} is ligated between two histidine residues and an aspartic acid side chain. The β-grasp domain is also very common in nature and is found in many proteins. For example, the β-grasp domain of streptococcal immunoglobulin binding proteins binds to IgG to avert immune responses. The affinity of these β-grasp domains is consistent and is utilized in streptococcal pathogenesis.

2.4.2 MHC-II α- Chain Binding

Out of all the superantigens of *S. pyogenes*, only SpeA and SSA have been seen to bind to MHC-II α chain. They generally bind with a very low affinity, varying in range of 0.1–1 μM (Redpath et al. 1999). There are many structures available for SAg/MHC-II complexes (Fig. 8). Loops that are formed above and below the OB fold have both polar and hydrophobic interactions amidst the residues flanking the peptide-binding groove of MHC-II α chain. The binding of SAgs to MHC-II α chain has also been experimentally explained by precise structures of SpeA in complex with mouse Vβ8.2 (Sharma et al. 2014). Participating residues of SpeA lay between the two domains and on the N-terminal α helix. CDR2 and HV4 loops are generally involved from the TCR. Additionally, the buried interphase between SAg and TCR is relatively smaller (540 $Å^2$). On studying the functional complexes of SpeA/TCR, one can construct a picture of circular, ternary complex wherein each component interacts with other. Here SAg hinders the normal MHC-II/TCR interaction through the N-terminal OB fold inserted between the MHC-II α chain and TCR β-chain. Hence, the TCR now connects both SAg and β-chain of MHC-II. It is contemplated that the coordinated interactions between TCR, MHC-II and SAg in a complete complex may make up for the low affinities seen (Redpath et al. 1999).

2.4.3 MHC-II β- Chain Binding

Rest of the SAgs (SpeC, SpeG, SpeH, SpeI, SpeJ, SpeK, SpeL, SpeM, SmeZ) bind to MHC-II via the β-chain. The structure of these SAg/TCR complexes is exemplified by SpeC (Fig. 8). Within the center of the binding junction, a Zn^{2+} ion is tetrahedrally coordinated by three ligands from SAg and one from the Histidine of MHC-II β-chain (His 81). As already mentioned, the binding of SAg to MHC-II β-chain is Zinc-dependent and their T-cell activating property is eliminated by addition of chelating agents like EDTA (Tripp et al. 2003). In comparison to

MHC-II α chain binding, the affinity of β-chain varies in range of 0.1–100 nM, which is three-fold higher (Proft et al. 1999). The binding of SpeC to β-chain of MHC-II leads to the flattening of the trio-complex (MHC-II/SpeC/TCR) as SpeC now binds to MHC-II at a distance (Sundberg et al. 2002). SpeC binds to TCR in more elaborate way; interacting with CDR1, CDR2, and CDR3 loops of β-chain of TCR, along with HV4/FR3 (Sundberg et al. 2002). The SpeC/TCR junction has 9 hydrogen bonds formed by side chains from SpeC and TCR. Even though some staphylococcal superantigens (like SEB) interact through these hydrogen bonds, the interaction of SpeC is highly definitive. This is because of unique conformational changes of CDR loops of human Vβ. The affinity of SpeC/TCR is very high, which can be emulated by the surface area of interaction between the complexes (810 Å^2). The high affinity of SpeC/MHC and SpeC/TCR may be requisite as of the linear arrangement of the MHC/SpeC/TCR ternary complex.

2.4.4   SAg Oligomerization

The aspect of SAg Oligomerisation is controversial. Almost all the superantigens are seem to work as monomers, except SpeC and SpeJ that are thought to oligo-merize and form dimers; although at higher concentrations. Crystal structure of SpeA has also proposed some oligomeric associations but the significance of such is still unclear. These alliances are generally weak and dependent a lot on the local concentration at the site of infection. SpeC was observed as both monomeric and dimeric form at neutral or alkaline pH (Li et al. 1997). Even though the exact reason for oligomerization is unknown, it is thought to play a role in MHC-II cross-linking and co-stimulation of T-cells. The crystal structure of SpeJ, on the other hand, showed a different way of homo-dimerization by TCR binding surface (Baker et al. 2004). It proposed two possible functions: MHC-II binding and cross-linking of MHC-II molecules. Recently SSA was seen as dimers in S. pyogenes supernatants (De Marzí et al. 2004). Intermolecular bridging was possible through cysteine (Cys26) linking. There are still a lot of open questions in regard to oligomerization and its function in pathogenicity.

2.4.5   Cross-Linking of MHC-II

Out of the streptococcal superantigens, only SpeC is seen to cross-link the MHC-II molecules at the antigen presenting cell surface. SpeC has been shown to cross-link MHC-II by formation of dimers through OB fold (Roussel et al. 1997). The dimerization of SpeC leaves the C-terminal β-grasp domain so as to bind to MHC-II in a zinc-mediated interaction (Li et al. 2001). This leads to formation of a complex (MHC/SpeC/TCR)$_2$. A second form of dimerization is shown by SpeJ, but in this case, the TCR binding site overlaps the dimerization site. Hence, the dimerization and cross-linking of MHC-II is mutually exclusive.

2.5   Association of SAg with Disease

Although the exact pathogenic function of superantigen is not known, it is still clear that these proteins imbalance the host immune system by activation of nonspecific T-cells and burst release of cytokines. Interestingly, the antibodies generated by a SAg are not expected to cross react with other superantigen, thus restricting the host efficiency to counteract multiple toxins (Bavari et al. 1999). Understanding the co-relation of superantigens with diseases is further more complicated because of variations in superantigen pattern present in the bacteria and their expression in vivo. Some of the superantigens are present in specific serotype while others or not. Also, some superantigens are expressed in earlier phase (low cell density; for example SpeJ) while others are expressed in later phase (high cell density; like SpeA, SmeZ). The disease manifestation is also dependent on the host genetics, especially the HLA haplotype. Understanding the association of any superantigen can be done in one of the three ways: (1) epidemiological studies (2) Presence of antibodies against these superantigens in sera (3) Restricted Vβ proliferation of T-cell. Below are some of the major outcomes of superantigens in diseases.

2.5.1   Invasive Diseases and Toxic Shock Syndrome (TSS)

The best example for studying the role of superantigens in diseases is Toxic Shock Syndrome, which could be either caused by Staphylococcus (StaphTSS) or Streptococcus (StrepTSS). StrepTSS is often related to spreading or disseminating bacteria, tissue infection, severe pain and bacteremia (Proft et al. 2003a, b). In acute stages, it can lead to multiple organ failure, destruction of soft tissue, shock and necrotizing soft tissue infections. These reactions sum up to a mortality rate as high as 30 %. Pyrogenic superantigens have been thought to be involved in mediating pathogenesis of StrepTSS. Seven of the superantigens (SpeA, SpeC, SpeG, SpeJ, SpeK, SSA and SmeZ) have been seen to be involved in pathogenesis of StrepTSS.

SpeA: Many epidemiological studies suggested the association of SpeA to StrepTSS (Jing 2006; Rogers et al. 2007). Interestingly, Eriksson et al. suggested that the risk factor for development of StrepTSS was the presence of SpeA rather than the clone M1T1 (Eriksson 1999). SpeA2 (found in M1) and SpeA3 (found in M3) were seen associated to reappearance of invasive *S. pyogenes* infections, as these were found in approximately 50 % of invasive infections and about two-third of StrepTSS cases (Talkington 1993; Musser 1991). On evolutionary terms, SpeA2 and SpeA3 are more recent and have possibly emerged through natural selection of isolates with high virulence. Contrarily, no association of SpeA was found with StrepTSS an invasive disease in many other studies (Mylvaganam 2000; Descheemaeker 2000).

SpeC: Most evidences suggest that there is no/inverse relation between presence of SpeC and development of invasive diseases and StrepTSS. SpeC-positive strains were mostly prevalent in StrepTSS cases collected in Europe and Canada (Lamagni et al. 2005, 2008; Barnham et al. 2002). Study of DelVecchio clearly showed a significantly higher prevalence of SpeC in Sydney that Northern territory, which was contrary to number of cases of invasive diseases and STSS which were more in Northern Territory (DelVecchio 2002).

SpeG and SpeJ: SpeG is a chromosomally inherited gene and thus is present in all the isolates. But it has been found to be more widespread in noninvasive cases than invasive ones (Murakami 2002). Nonetheless, most of the studies across the world suggest no correlation of SpeG with invasive disease and TSS (Descheemaeker 2000). Similarly, SpeJ was thought to be not related to invasive diseases and TSS but recent study in Portugal seem to find an association (Friaes 2012).

SpeK and SSA: It is striking to see the association of M3T3 strains with StrepTSS in mid-1980s, after the acquisition of phage bearing SpeK (Ikebe et al. 2002). Similarly, SSA was acquired in M3 clone in 1937 (Reda 1994) and was seen to be associated with invasive diseases in Japan (Murakami 2002), in contrast to studies in Europe that found its association with pharyngitis (Friaes 2012).

SmeZ: Along with SpeA, SmeZ has been seen to play a major role in StrepTSS (Proft et al. 2003a, b). Even though studies have shown association of SmeZ in disease manifestation, no study has been made to scrutinize role of different alleles of SmeZ in pathogenesis of StrepTSS. Recently Turner et al. showed that the super-antigenic activity of SmeZ was abolished by a conserved, naturally mutation in M3 isolate (Turner et al. 2012).

2.5.2 SAg and Autoimmunity

Recurrent infections of *S. pyogenes* can lead to autoimmune sequelae like Rheumatic heart disease. It is thought that bacteria producing superantigens can initiate autoimmunity in vulnerable hosts (Delogu et al. 2011). Several mechanisms have been proposed as of how autoimmunity is triggered. Molecular Mimicry is one of the most studied mechanisms where epitopes are shared between host and pathogen that leads to production of antibodies against self. Superantigens are known to effect T-cell proliferation and production of cytokine response. It is known that 'self-reactive' T-cells circulate in blood of a healthy individual. Generally, these T-cells are anergetic and present in low-abundance, but can activate and potentially cause autoimmunity if stimulated beyond the threshold levels. Hence, if any superantigen activates such 'self-reactive' T-cell, it can lead to massive proliferation of such T-cells that can potentiate autoimmunity (Hirota et al. 2007). Beside this, the binding to superantigen to MHC-II can activate B-cells, macrophages, and thus secrete enormous amount of cytokines and co-stimulatory molecules that can augment tissue damage. These reactions may contribute in aberrant display of self-proteins that can stimulate the 'self-reactive' T-cells. Many of the autoimmune diseases like rheumatic heart disease (Ellis et al. 2005) and rheumatic arthritis (Sottini et al. 1991) have been linked to superantigens.

Rheumatic Fever

There has been a substantial correlation between M18 isolates of acute rheumatic fever with SpeL and SpeM. It was found that all the isolates (M18) carried these two SAgs over a period of 69 years. Also, antibodies against SpeL and SpeM were more common in sera of these patients (Smoot et al. 2002). Similarly, SpeK has been associated with M89 isolates of rheumatic fever (RF) in New Zealand (Proft et al. 2003a, b). Interestingly, it was seen that the CD4$^+$ and CD8$^+$ cells of rheumatic heard disease (RHD) patients respond differently to SpeA. In vitro sub-culturing of SpeA with CD4$^+$ cells of RF patients showed Th1 response that produced IL-2 while CD4$^+$ cells of RHD patients produced enormous amount of Il-4 and IL-10 (TH2 cytokine profile) (Bhatnagar et al. 1999).

Kawasaki Disease

Kawasaki Disease is an autoimmune disease affecting the blood vessels. It is an acute vasculitic disease accompanied by fever and can lead to damage of coronary arteries. Recent studies have shown biasing of Vβ subtypes of T-cells in patients of Kawasaki disease. Precisely, Vβ2 T-cells were involved, thus letting the focus on SpeC (Yoshioka 1999). Later on, Yoshioka showed higher levels of antibodies against SpeC in sera of patients with acute Kawasaki disease (Yoshioka 2003). Matsubara et al. also supported the hypothesis of the role of SAg in Kawasaki disease by demonstrating the presence of IgM against SpeC, SpeA, and SpeM during initial phase of disease (Matsubara et al. 2006). Contrarily, some epidemiological data could not find any correlation of superantigen to Kawasaki disease (Nomura 2002; Gupta-Malhotra 2004).

Psoriasis

Psoriasis is also one of the immune-mediated diseases that cause skin lesions including papules and plaques that are generally itchy. Leung showed skewing of Vβ2 T-cells that were indicative of role of SpeC in psoriasis. Interestingly in his study all the selected isolated were positive for SpeC (Leung 1993). On the other side, Travers raised an alternative hypothesis by showing that host genetic factors (HLA type; HLA-DR) were too involved in the hypothesis. They demonstrated that the HLA-DR positive keratinocytes were directly interacting with SAgs (Travers 1999).

2.5.3   Other Diseases

Superantigens have been associated with other diseases too. Scarlet fever, for instance, was initially described as being caused by Scarlatina toxin in 1924 (Dick and Dick 1924). Later it was identified that the 'rashes' were due to SpeA and SpeC.

Superantigens are also seen associated with ulcerative colitis and chronic sinusitis/nasal polyposis (Shiobara 2007; Tripathi 2004). Shiobara demonstrated the activation and proliferation of Vβ4 T-cells, indicating the role of SmeZ2 on the disease manifestation (Shiobara 2007). Whereas Tripathi showed the correlation of IgE specific to SAgs and chronic sinusitis/nasal polyposis. These specific IgE antibodies were markedly increased in patients (7 out of 21) with chronic sinusitis/nasal polyposis as compared to controls (0 out of 13) (Tripathi 2004). Superantigens have also been seen to be one of the causative agents in narcolepsy (Fontana 2010).

2.6 Host Factors: Role of HLA Type

Interestingly, mouse escalates to relatively lower T-cell response to superantigens as compared to humans. It is thought to be due to differences in sequences of mouse and human MHC-II, which leads to low binding of SAg to mouse MHC-II (Sriskandan et al. 2001). Additionally, there are also differences within the human MHC-II (HLA class II) can lead to differential presentation and activation of individual SAgs. For instance, SpeA preferentially binds to HLA-DQ rather than HLA-DR (Imanishi et al. 1992; Norrby-Teglund et al. 2002). During the initial characterization of superantigens, it was indicated that this effect was due to defiance of MHC class restriction. This focused the scientific community on similarities in SAg binding instead of their differences. Whereas SAgs like SpeC were using HLA-DR (DR4) more efficiently than HLA-DQ or HLA-DP in T-cell activation, SpeA that preferentially uses HLA-DQ (Llewelyn et al. 2004; Norrby-Teglund et al. 2002). But, recent report by Kotb et al. depicted the correlation of HLA haplotype to the susceptibility to SAg associated diseases (Kotb et al. 2002). Since all the patients in this study were essentially infected with different *S. pyogenes* strains (each possessing a different set of SAgs), it is difficult to relate HLA association of each SAg. It is hence, challenging to find molecular mechanisms that correlate SAg/HLA class II/TCR interactions. Different studies have also showed co-relation of HLA with susceptibility of diseases. Some patients with a particular HLA haplotype are more susceptible in having certain SAg related disease than other. For instance, patients with the DRB1*1501/DQB1*0602 haplotype showed significantly reduced responses and were less likely to develop severe systemic disease (Kotb et al. 2002). DRB1*03/DQB1*0201 haplotype was associated with protection from necrotizing fasciitis (Kotb et al. 2002). Interestingly, where DRB1*1501/DQB1*0602 was seen to be protective against StrepTSS, DRB1*14/DQB1*0503 were found to be predisposal to STSS (Normark and Normark 2002). Similar study showed that the allele DRB3*01:01:02:01 had a positive association with RHD, whereas the DQB1 loci alleles did not show any significant association.

2.7 Evolutionary Paths of Streptococcal and Staphylococcal SAgs

The most studied superantigens are those produced by *S. pyogenes* and *Staphylococcus aureus*. *S. aureus* secrete more than 14 different SAgs, known as Staphylococcal Exotoxins (SEs) A, B, C1-3, D, E, G, H, I, J, K,L, and M and toxic shock syndrome toxin -1 (TSST-1) (Marrack and Kappler 1990). Similarly, streptococcal superantigens are named as streptococcal pyrogenic exotoxins (SPEs) and are 11 in number (A, C, G, H, I, J, K, L, M, SSA, and SmeZ). These are just the conventional superantigens as there is an increasing number of superantigens as and their alleles being discovered after the whole genome sequencing. As already stated, SAgs show variations at primary nucleotide as well as amino acid sequence level. While some of the superantigens like SEA and SEE are about 83 % similar in primary amino acid level, other superantigens like SEH and SEC1 are only similarity up to 20 %. An approximate of 15 % of residues is always conserved in all the known superantigens. Most of the 'conserved' residues are either centrally localized or are at the C-terminal. Interestingly, some proteins with high structural similarity to SAgs have been recently found in *S. aureus* and are named as Staphylococcal superantigen-like protein (SSLs), but they do not possess functional properties of SAgs.

Fig. 9 Family tree of staphylococcal and streptococcal superantigen. The tree was generated by Clustal (Thompson et al. 1994)

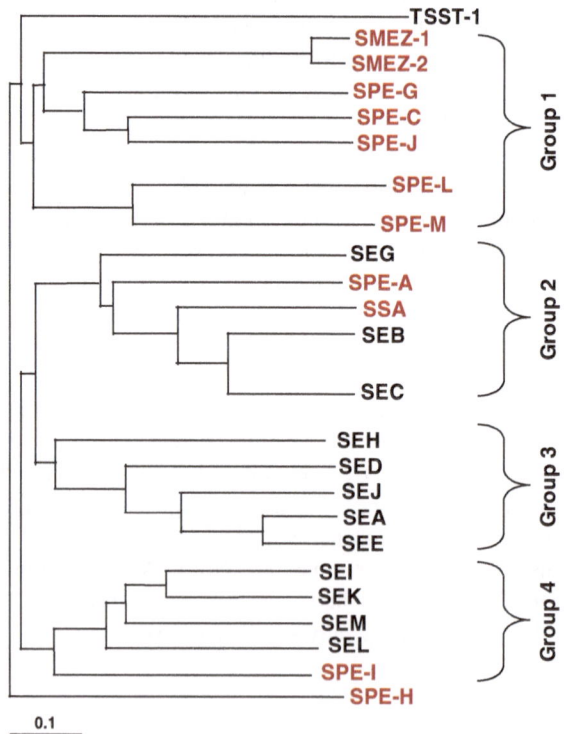

On aligning all the SAgs based on their amino acid sequences, one can divide them into four distinct groups (Fig. 9) (Proft et al. 2000). Group 1 includes SmeZ (SmeZ-1 and SmeZ-2), SpeG, SpeC, SpeJ, SpeL, and SpeM. Group 2 includes SEG, SpeA, SSA, SEB, and SEC. Group 3 has SEH, SED, SEJ, SEA, and SEE whereas, Group 4 includes SEI, SEK, SEM, SEL, and SpeI. TSST-1 and SpeH remain as outliers as they do not classify into any of the group even though they have homology of around 20–30 % with other superantigens. It can be seen that streptococcal SAgs like SpeA, SpeH, SpeI and SSA are most closely related to *S. aureus* SAgs than any other SAg. Since all of these are bacteriophage related, it could be possible that they were horizontal transferred from *S. aureus* instead of deriving from an existing streptococcal SAg gene. The streptococcal and staphylococcal superantigens are proteins that range from 22 to 28 kDa with an average of 194–242 amino acids in length (Alouf and Müller-Alouf 2003). Although the similarity between SAgs is as low as 20 % in some cases, they embrace an exceptionally similar 3D-structure.

3 Therapeutic Strategies to Combat SAg

Generally, in diseases like StrepTSS, time is a crucial factor in treatment. Early diagnosis could help in recovering of patient through proper treatment and appropriate antibiotics. However, antibiotics may not work once the disease becomes acute. Eradicating the bacteria might not help as the detrimental effect by superantigens take a toll. Thus, in such cases, superantigens need to be either neutralized or made dysfunctional so as to reduce its effect. However, bacteria-host interactions and its pathogenesis should be emphasized during development of therapeutic strategies. The destructive effect of superantigens can be impeded by one of the three ways: (1) downregulation of superantigen production, (2) super-antigen neutralization, and (3) downregulation of SAg-induced inflammation. Down regulation of SAg production seems to be the simplest method but indeed is the most difficult to follow up. There could be though, many ways to do so. Antisense- DNA or siRNA-mediated downregulation is one of the ways, wherein infected patients are treated with it. Even though this is a very specific targeting method, it has many problems including passage of these molecules from one bacterial cell to another. Other ways of superantigen impediment are its neutralization or downregulation of inflammation caused by SAg. Once secreted, SAgs can be neutralized by blocking peptides or specific antibodies. Both methods have been explained in details below.

3.1 Intravenous Immunoglobulins (IVIG)

Intravenous immunoglobulins (IVIG) are a sterile preparation of concentrated antibodies from pooled plasma/serum samples of at least 1000 healthy donors. Even though IVIG is dominated by IgG, there might be few traces of IgA found as well.

Since a lack of protective antibodies is thought to be one of the risk factors in development of diseases like StrepTSS and necrotizing fasciitis, IVIG could be one of the potential collateral therapies (Basma et al. 1999). Since, IVIG is a pool of all the antibodies derived from thousands of healthy doors, it is nonspecific and thus can cover various serotypes of *S. pyogenes* as well as a complete spectrum of superantigens and other exotoxins. IVIG is commonly used in as a therapy in many immune deficiencies (like hypogammaglobulinemia, X-linked agammaglobulinemia, etc.), low antibody titers, autoimmune diseases, and inflammatory diseases (like Kawasaki disease) (Katz et al. 2011).

3.1.1 Mechanism of Action of IVIG

Several different mechanistic actions of IVIG have been ascribed that contribute to its positive effect in streptococcal diseases. These comprise blockage of Fc receptor, modulation of cytokine responses, causing variation in immune cell functioning, interaction with idiotype–anti-idiotypic network and antigen neutralization (Mouthon et al. 1996). With respect to SAg-mediated diseases, like StrepTSS, IVIG is seen to neutralize antigens (including SAgs), mediate bacterial opsonization, and modulate cytokine responses. IVIG contains antibodies that can opsonize the bacteria and hence lead to its clearance by phagocytic killing (Weisman et al. 1994; Basma et al. 1998). Antibodies against M-protein have also been found in IVIG preparations (Basma et al. 1998). On analyzing plasma samples from the patients, pre- and post-IVIG treatment could show a significant increase in anti-M protein antibodies in the samples (Basma et al. 1998). Hence, bacterial opsonization and then clearance is likely to be one of the mechanistic actions of IVIG, thus contributing to its effectiveness as an adjunctive therapy. Contrarily, when efficacy of IVIG was checked, in murine models of necrotizing fasciitis, along with clindamycin and penicillin, no beneficial effect IVIG was observed, thus failing to support the previous hypothesis (Patel et al. 2000). Comparing these three treatments, IVIG did not augment bacterial clearance. But it could be anticipated that human antibodies were not efficient opsonins for mouse phagocytes.

Another hypothesis states the present M-protein to be protective, inhibiting the M-protein-mediated activation of neutrophils; but it still needs to be tested. IVIG also contains potent neutralizing antibodies against the streptococcal and staphylococcal superantigens and are seen to inactivate them. These antibodies help in inhibiting the massive proliferation and cytokine releases under in vitro conditions (Darenberg et al. 2004). Even though IVIG was able to neutralize all the SAgs, the degree of inhibition varied among different SAgs. Also, IVIG was more effective in inhibiting streptococcal rather than staphylococcal superantigens (Darenberg et al. 2004). All these studies demonstrated that, at physiological concentration, all the superantigens are inhibited up to 100 % by the high dose of administered IVIG (1–2 g/kg body weight). Also, IVIG can be considered as a 'cocktail' of different antibodies against various virulence factors of *S. pyogenes*, and thus can be provided to the patient. Additionally, anti-DNaseB and anti-streptolysin O antibodies are found in IVIG that can block the activity of these important virulence factors.

As already mentioned, IVIG is primarily composed of IgG, but can also have varying amounts of IgA and/or IgM. However, using such IVIG preparations showed an inhibitory effect on opsonizing and toxin-neutralizing capability of IVIG (Norrby-Teglund et al. 2000). IgA and IgM were inhibiting the superantigens, especially SpeA, where most efficient neutralization was done by a mixture containing all of the three immunoglobulins (IgG, IgA, and IgM) (Norrby-Teglund et al. 2000). Overall, optimization of IVIG therapy still remains to be proved in a clinical setting.

3.1.2 Variation of Cytokine Responses

IVIG is seen to modulate cytokine responses of the host by direct antigen neutralization, Fc interactions, soluble immune components, and induction of regulatory cytokines (Mouthon et al. 1996). Co-culturing human monocytes with IVIG resulted in strong induction of IL-1ra (InterLeukin-1 receptor antagonist), serving as an antagonist to production and signaling of IL-1 (Poutsiaka et al. 1991). Similarly, response of IL-8 is upregulated during co-culturing human monocytes with IVIG (Ruiz De Souza et al. 1995). IL-1 is a proinflammatory cytokine; hence production of its antagonist leads to decrease in inflammation. Even though, IL-8 too is a proinflammatory cytokine, due to its chemotactic properties, it is thought that release of IL-8 systemically could have an anti-inflammatory effect by declining the amount of neutrophils at the site of inflammation. In vivo, IVIG is sown to be a strong inhibitor of superantigen-induced T-cell proliferation leading to chemokine and cytokine production. It is seen to strongly repress production of Th1 cytokine IFNγ and TNFβ (Darenberg et al. 2004; Andersson et al. 1994). This effect, although lower, was also seen when IVIG was added to SAg after 24 h of stimulation (Andersson et al. 1994). Interestingly, on addition of IVIG, Andersson et al. could see can 'two-way' effect on SAg: by upregulation of IL-8 and downregulation of IL-6 (Andersson et al. 1994). There were conflicting results seen for the effect of IVI on IL-1 with one study showing no effect (Skansén-Saphir et al. 1994) while the other demonstrating significantly lower amounts of IL-1 (Norrby-Teglund et al. 2000). Hence, the effect of IVIG on cytokine response still needs to be explicated. It has to be noted that the major type of cytokine (Th1) is strongly and completely inhibited by IVIG. This might represent one of the major mechanisms of IVIG's clinical efficacy. Modulation of cytokine response by IVIG has been seen to work in vivo in many diseases like sever *S. pyogenes* infections (reduction of TNFα and IL-6) and Kawasaki patients (upregulation of IL-1ra and IL-8).

3.1.3 Clinical Studies

There are many clinical cases wherein IVIG was used as an adjunctive therapy in streptococcal cases, especially in StrepTSS (Jolles et al. 2005). Many case reports (Schrage et al. 2006; Barry et al. 1992; Cawley et al. 1999), observational cohorts

(Norrby-Teglund et al. 2005), case-control study (Kaul et al. 1995) and multi-centered placebo-control study (Darenberg et al. 2003). The multi-centered placebo-control study was prematurely terminated because of slow patient recruitment (Darenberg et al. 2003). StrepTSS is generally seen along with necrotizing fasciitis, thence making the mortality even higher. The major factor in such a high mortality rate associated with these diseases is the time factor: the time taken in diagnosis. There is a need of early surgery to reduce inflammation and restrict the spread of local infection.

Recently, another clinical study called INSTINCT (Immunoglobulin for Necrotizing Soft Tissue Infection: a Randomized Control Trial) has been initiated by Rigshospitalet in Denmark that would investigate the effect of IVIG in a double blind study, taking 1:1 patients treated with IVIG and placebo (https://www.clinicaltrials.gov/ct2/show/study/NCT02111161). They are currently in process of patient recruitment (April 2014).

The data suggests that IVIG is beneficial as a collateral therapy and is effective against wide varieties of strains and serotypes. It has been seen to neutralize superantigens, modulate of Fc receptor blockade and expression, inhibit membrane attack complexes (C5b-9), activate complement system, and facilitate opsonization of GAS bacteria. Clinical studies' including cohort studies and case-control studies definitely substantiates the beneficial effect of administration of IVIG to patients of StrepTSS and necrotizing Fasciitis. Even though efficacy of IVIG during StaphTSS still needs to be validated, a few case studies hint to its importance in some severe cases.

3.2 Pyrogenic Superantigens as Cures for Human Diseases

In 1990, Dohlsten et al. showed that the addition of staphylococcal superantigen SEA changed its target specificity in T-cell cytotoxicity assay (Dohlsten et al. 1990). Later, they investigated the role of SEA as an immune-regulator in vivo (Hedlund et al. 1993). They found that the superantigen could activate and guide T-cells to eliminate MHC-II expressing cells. This was then named as superantigen-dependent cellular cytotoxicity (SDCC). Even though T-cell subsets could arbitrate SDCC, NK cells were unable to do so. It could be due to absence of receptors for superantigens on NK cells (Lando et al. 1991). Hence, SDCC was proposed to be one of the ways for immune evasion and bacterial survival during infection (Kalland et al. 1991). Interestingly, this phenomenon was not only limited to staphylococcal superantigens but was also true for streptococcal ones. Thence it was later used as a tool in elimination of MHC class II expressing tumor cells (Patterson et al. 2014). Since SDCC is involved in killing of MHC class II cells, it was used for eradication of MHC class II tumors. Target cell specificity could also be achieved by conjugating superantigens to a suitable monoclonal antibody that is specific for a particular tumor antigen (Kalland et al. 1993). This novel approach, also known as SAg directed T-cell killing, may substantiate new approaches for therapeutics in cancer immunology.

Another way superantigen could be 'good' for us, is in treatment of autoimmunity. Naïve T-cells respond and become activated by SAgs. These T-cells are later anergized or deleted from the host system. But, other T-cells that are activated by some other antigen at the time of SAg activation are not anergized. This could be of well use as therapeutics for development of specific T-cell response against any specific antigen (Torres et al. 2001).

3.3 Why Bacteria Produce SAg

Although many microorganisms produce proteins/chemical substances that can be beneficial to us, either by being an antibiotic or fermented product. But, *S. pyogenes* is a strict human pathogen and obviously did not produce superantigens for any benefit to the host. Production of superantigens is solely being damaging to the host as these molecules are persuasive inducers of T-cells. Interestingly, these molecules give an edge to the pathogen by allowing it to circumvent immune response of the host, thus "taking over". Superantigens are seen to deter the immune system by one of the three mechanisms.

First, its capability to induce SDCC (Superantigen-Dependent Cellular Cytotoxicity) that eradicates cells hindering SAg-activated T-cell proliferation. Production of IL-8 and other inflammatory cytokines worsens the situation for the host, thus enhancing SDCC. Eradication of MHC-II expressing cells may also lead to anergy and further incapability of T-cell stimulation. Second, massive production of cytokines and proinflammatory chemokines may lead to specific apoptosis of T-cells carrying definite Vβ binding site for that Superantigen. Lastly, activation of nonspecific T-cells helps in "fooling" the host, thus diverting the attention from a centered and specific immune response toward the pathogen. But it should be kept in mind that functions of superantigen differ significantly on the host genetics and the interactions between the bacteria and host immune system. Therefore, a particular superantigen might be generating different responses in different hosts. There is a need to study superantigens and the host simultaneously to understand the importance of these proteins in pathogenesis of *S. pyogenes*.

Understanding superantigens has given us an insight to the immune modulations and interactions that can occur between a host and pathogen. It can hence be anticipated that exploration of these proteins would help us, in the near future, toward therapeutic benefits from these subset of proteins.

3.4 Concluding Remarks

Indeed, superantigens are an appealing group of proteins that has fascinated a lot of scientists. Studying these proteins for their function and structure along with its mechanism of action could help us pave new ways in revealing new pathogenic mechanism and may have therapeutic benefits. It must be kept in mind that even if every superantigen has a similar structure and function, they vary significantly based on host genetics and environmental factors. Studying these proteins has enlightened us to new pathogenetic mechanisms of bacteria in modulating and evading host immune response. But, a lot of it is still unexplored and need attention. It is quite possible that these proteins might have any additional function or any hidden potential that still needs to be investigated. One should be quite cautious in interpreting these molecules due to huge variations in responses. Studying these proteins has unleased many therapeutic ways and is anticipated to do so in coming future.

References

Agniswamy, J., Lei, B., Musser, J. M., & Sun, P. D. (2004). Insight of host immune evasion mediated by two variants of group a Streptococcus Mac protein. *Journal of Biological Chemistry, 279*(50), 52789–52796.

Alouf, J. E., & Müller-Alouf, H. (2003). Staphylococcal and streptococcal superantigens: Molecular, biological and clinical aspects. *International Journal of Medical Microbiology (IJMM), 292*(7–8), 429–440. doi:10.1078/1438-4221-00232.

Alouf, J. E., Mueller-Alouf, H., & Koehler, W. (1999). Superantigenic Streptococcus pyogenes erythrogenic/pyrogenic exotoxins. In J. E. Alouf & J. H. Freer (Eds.), *Sourcebook of bacterial protein toxins*. San Diego: Academic Press

Andersson, U., Björk, L., Skansén-Saphir, U., & Andersson, J. (1994). Pooled human IgG modulates cytokine production in lymphocytes and monocytes. *Immunological Reviews, 139* (139), 21–42.

Andrews R. M., McCarthy J., Carapetis J. R., Currie B. J. (2009). Skin disorders, including pyoderma, scabies, and tinea infection. *Pediatric Clinics of North America, 56*(6), 1421–40.

Ato, M., Ikebe, T., Kawabata, H., Takemori, T., & Watanabe, H. (2008). Incompetence of neutrophils to invasive group a Streptococcus is attributed to induction of plural virulence factors by dysfunction of a regulator. *PLoS ONE, 3*, e3455.

Andreoni, F., Ogawa, T., Ogawa, M., Madon, J., Uchiyama, S., Schuepbach, R. A., & Zinkernagel, A. S. (2014). The IL-8 protease SpyCEP is detrimental for group A Streptococcus host-cells interaction and biofilm formation. *Frontiers in Microbiology, 5*, 339. doi:10.3389/fmicb.2014.00339.

Baker, H. M., Proft, T., Webb, P. D., Arcus, V. L., Fraser, J. D., & Baker, E. N. (2004). Crystallographic and mutational data show that the streptococcal pyrogenic exotoxin J can use a common binding surface for T-cell receptor binding and dimerization. *Journal of Biological Chemistry, 279*(37), 38571–38576. doi:10.1074/jbc.M406695200.

Baker, M. D., & Acharya, K. R. (2004). Superantigens: Structure-function relationships. *International Journal of Medical Microbiology, 293*, 529–537.

Baker, M., Gutman, D. M., Papageorgiou, A. C., Collins, C. M., & Acharya, K. R. (2001). Structural features of a zinc binding site in the superantigen strepococcal pyrogenic exotoxin A (SpeA1): Implications for MHC class II recognition. *Protein Science: A Publication of the Protein Society, 10*(6), 1268–1273. doi:10.1110/ps.330101.

Barnham, M. R. D., Weightman, N. C., Anderson, A W., & Tanna, A. (2002). Streptococcal toxic shock syndrome: a description of 14 cases from North Yorkshire, UK. *Clinical Microbiology and Infection: The Official Publication of the European Society of Clinical Microbiology and Infectious Diseases, 8*(3), 174–181. doi:10.1046/j.1469-0691.2002.00396.x.

Barry, W., Hudgins, L., Donta, S. T., & Pesanti, E. L. (1992). Intravenous Immunoglobulin therapy for toxic shock syndrome. *JAMA, 267*(24), 3315–3316. doi:10.1001/jama.1992.03480240077038.

Basma, H., Norrby-Teglund, A., Geer, A. M. C., Low, D. E., El-Ahmedy, O., Dale, J. B., et al. (1998). Opsonic antibodies to the surface M protein of group A streptococci in pooled normal immunoglobulins (IVIG): Potential impact on the clinical efficacy of IVIG therapy for Severe Invasive Group A Streptococcal Infections. *Infection and Immunity, 66*(5), 2279–2283.

Basma, H., Norrby-Teglund, A., Guedez, Y., Geer, A. M. C., Low, D. E., El-ahmedy, O., et al. (1999). Risk factors in the pathogenesis of invasive group a streptococcal infections: Role of protective humoral immunity. *Infection and Immunity, 67*(4), 1871–1877.

Bavari, S., Ulrich, R. G., & LeClaire, R. D. (1999). Cross-reactive antibodies prevent the lethal effects of Staphylococcus aureus superantigens. *Journal of Infectious Diseases, 180*, 1365–1369.

Bhatia, S., Edidin, M., Almo, S. C., & Nathenson, S. (2006). B7-1 and B7-2: Similar costimulatory ligands with different biochemical, oligomeric and signaling properties. *Immunology Letters, 104*, 70–75.

Bhatnagar, A., Grover, A., & Ganguly, N. K. (1999). Superantigen-induced T cell responses in acute rheumatic fever and chronic: Rheumatic heart disease patients. *Clinical and Experimental Immunology, 116*, 100–106. doi:10.1046/j.1365-2249.1999.00853.x.

Bisno, A. L., Brito, M. O., & Collins, C. M. (2003). Molecular basis of group A streptococcal virulence. *Lancet Infectious Diseases, 3*(4), 191–200.

Bohach, G. A., Hauser, A. R., & Schlievert, P. M. (1988). Cloning of the gene, speB, for streptococcal pyrogenic exotoxin type B in Escherichia coli. *Infection and Immunity, 56*(6), 1665–1667.

Bonnetblanc, J. M., & Bédane, C. (2003). Erysipelas: Recognition and management. *American Journal of Clinical Dermatology,*. doi:10.2165/00128071-200304030-00002.

Brinkmann, V., et al. (2004). Neutrophil extracellular traps kill bacteria. *Science, 303*, 1532–1535.

Brosnahan, A. J., Mantz, M. J., Squier, C. A., Peterson, M. L., & Schlievert, P. M. (2009). Cytolysins augment superantigen penetration of stratified mucosa. *Journal of Immunology, 182*, 2364–2373.

Bryant, A. E., Bayer, C. R., Chen, R. Y. Z., Guth, P. H., Wallace, R. J., & Stevens, D. L. (2005). Vascular dysfunction and ischemic destruction of tissue in Streptococcus pyogenes infection: the role of streptolysin O-induced platelet/neutrophil complexes. *The Journal of Infectious Diseases, 192*(6), 1014–1022. doi:10.1086/432729.

Buchanan, J. T., et al. (2006). DNase expression allows the pathogen group A Streptococcus to escape killing in neutrophil extracellular traps. *Current Biology, 16*, 396–400.

Carapetis, J. R., Steer, A. C., Mulholland, E. K., & Weber, M. (2005). The global burden of group A streptococcal diseases. *Lancet, 5*(November), 685–694.

Carroll, R. K., & Musser, J. M. (2011). From transcription to activation: how group a streptococcus, the flesh-eating pathogen, regulates SpeB cysteine protease production. *Molecular Microbiology, 81*(3), 588–601. doi:10.1111/j.1365-2958.2011.07709.x.

Cawley, M. J., Briggs, M., Haith, L. R., Reilly, K. J., Guilday, R. E., Braxton, G. R., & Patton, M. L. (1999). Intravenous immunoglobulin as adjunctive treatment for streptococcal toxic shock syndrome associated with necrotizing fasciitis: Case report and review. *Clinical And Translational Science, 19*(9), 1094–1098.

Chaussee, M. S., Liu, J., Stevens, D. L., & Ferretti, J. J. (1996). Genetic and phenotypic diversity among isolates of Streptococcus pyogenes from invasive infections. *The Journal of Infectious Diseases, 173*, 901–908.

Chiappini, N., Seubert, A., Telford, J. L., Grandi, G., Serruto, D., Margarit, I., et al. (2012). Streptococcus pyogenes SpyCEP influences host-pathogen interactions during infection in a murine air pouch model. *PLoS ONE, 7*, e40411.

Chong, B. F., Blank, L. M., Mclaughlin, R., & Nielsen, L. K. (2005). Microbial hyaluronic acid production. *Applied Microbiology and Biotechnology*. doi:10.1007/s00253-004-1774-4.

Cole, J. N., Barnett, T. C., Nizet, V., & Walker, M. J. (2011). Molecular insight into invasive group a streptococcal disease. *Nature Reviews Microbiology, 9*(10), 724–736. doi:10.1038/nrmicro2648.

Cole, J. N., et al. (2010). M protein and hyaluronic acid are essential for in vivo selection of covRS mutations characteristic of invasive M1T1 group a Streptococcus. *mBio, 1,* e00191–00110.

Commons, R. J., Smeesters, P. R., Proft, T., Fraser, J. D., Robins-Browne, R., & Curtis, N. (2014). Streptococcal superantigens: Categorization and clinical associations. *Trends in Molecular Medicine, 20*(1), 48–62. doi:10.1016/j.molmed.2013.10.004.

Courtney, H. S., Hasty, D. L., & Dale, J. B. (2002). Molecular mechanisms of adhesion, colonization, and invasion of group A streptococci. *Annals of Medicine 34,* 77–87.

Cremer, N., & Watson, D. (1960). Host-parasite factors in group a streptococcal infections: A comparative study of streptococcal pyrogenic toxins and gram-negative bacterial endotoxin. *The Journal of Experimental Medicine, 112*(6), 1037–1053.

Cunningham, M. W. (2000). Pathogenesis of group A streptococcal infections. *Clinical Microbiology Reviews, 13,* 470–511.

Dale, J. B., Penfound, T. A., Chiang, E. Y., Walton W. J. (2011). New 30-valent M protein-based vaccine evokes cross-opsonic antibodies against non-vaccine serotypes of group A streptococci. *Vaccine, 29*(46), 8175–8.

Dale, J. B., & Chiang, E. C. (1995). Intranasal immunization with recombinant group A streptococcal M protein fragment fused to the B subunit of Escherichia coli labile toxin protects mice against systemic challenge infections. *The Journal of Infectious Diseases, 171,* 1038–1041.

Dano, K., Andreasen, P. A., Grondahl-Hansen, J., Kristensen, P., Nielsen, L. S., & Skriver, L. (1985). Plasminogen activators, tissue degradation, and cancer. *Advances in Cancer Research, 44,* 139–266.

Darenberg, J., Ihendyane, N., Sjo, J., Aufwerber, E., Haidl, S., Follin, P., et al. (2003). Intravenous immunoglobulin G therapy in streptococcal toxic shock syndrome. *Clinical Infectious Diseases, 37,* 333–340

Darenberg, J., So, B., Normark, B. H., & Norrby-Teglund, A. (2004). Differences in Potency of Intravenous Polyspecific Immunoglobulin G against Streptococcal and Staphylococcal Superantigens : Implications for Therapy of Toxic Shock Syndrome. *Clinical Infectious Diseases, 38,* 836–842.

De Marzí, M. C., Fernández, M. M., Sundberg, E. J., Molinero, L., Zwirner, N. W., Llera, A. S., et al. (2004). Cloning, expression and interaction of human T-cell receptors with the bacterial superantigen SSA. *European Journal of Biochemistry/FEBS, 271*(20), 4075–83. doi:10.1111/j.1432-1033.2004.04345.x.

Delogu, L. G., Deidda, S., Delitala, G., & Manetti, R. (2011). Infectious diseases and autoimmunity. *Journal of Infection in Developing Countries.* doi:10.3855/jidc.2061.

DelVecchio, A. (2002). NAD-glycohydrolase production and speA and speC distribution in group a streptococcus (GAS) isolates do not correlate with severe GAS diseases in the Australian population. *Journal of Clinical Microbiology, 40,* 2642–2644.

Descheemaeker, P. (2000). Molecular characterisation of group A streptococci from invasive and non-invasive disease episodes in Belgium during 1993–1994. *Journal of Medical Microbiology, 49,* 467–471.

Deutscher, M., Lewis, M., Zell, E. R, Taylor, T. H., Jr, Van Beneden, C., Schrag, S. (2011). Incidence and severity of invasive Streptococcus pneumoniae, group A Streptococcus, and group B Streptococcus infections among pregnant and postpartum women. *Clinical Infectious Diseases, 53*(2), 114–123.

Dick, G. F., & Dick, G. H. (1924). The etioloy of scarlet fever. *JAMA, 82*(4), 301–302. doi:10.1001/jama.1924.02650300047013.

Dohlsten, M., Lando, P. A., Hedlund, G., Trowsdale, J., & Kalland, T. (1990). Targeting of human cytotoxic T lymphocytes to MHC class II-expressing cells by staphylococcal enterotoxins. *Immunology, 71*(1), 96–100.

Edwards, R. J., Taylor, G. W., Ferguson, M., Murray, S., Rendell, N., Wrigley, A., et al. (2005). Specific C-terminal cleavage and inactivation of interleukin-8 by invasive disease isolates of Streptococcus pyogenes. *Journal of Infectious Diseases, 192*, 783–790.

Ellis, N. M. J., Li, Y., Hildebrand, W., Fischetti, V. A., & Cunningham, M. W. (2005). T cell mimicry and epitope specificity of cross-reactive t cell clones from rheumatic heart disease. *The Journal of Immunology, 175*(8), 5448–5456. doi:10.4049/jimmunol.175.8.5448.

Eriksson, B. K. (1999). Invasive group A streptococcal infections: T1M1 isolates expressing pyrogenic exotoxins A and B in combination with selective lack of toxin-neutralizing antibodies are associated with increased risk of streptococcal toxic shock syndrome. *Journal of Infectious Diseases, 180*, 410–418.

Eurosurveillance. (2005). *Eurosurveillance.*

Fernie-King, B. A., et al. (2001). Streptococcal inhibitor of complement (SIC) inhibits the membrane attack complex by preventing uptake of C567 onto cell membranes. *Immunology, 103*, 390–398.

Ferretti, J. J., et al. (2001). Complete genome sequence of an M1 strain of Streptococcus pyogenes. *Proceedings of the National Academy of Sciences USA, 98*, 4658–4663.

Fontana, A. (2010). Narcolepsy: Autoimmunity, effector T cell activation due to infection, or T cell independent, major histocompatibility complex class II induced neuronal loss? *Brain, 133*, 1300–1311.

Fraser, J., Arcus, V., Kong, P., Baker, E., & Proft, T. (2000). Superantigens—powerful modifiers of the immune system. *Molecular Medicine, 6*, 125–132.

Friaes, A. (2012). Group A streptococci clones associated with invasive infections and pharyngitis in Portugal present differences in emm types, superantigen gene content and antimicrobial resistance. *BMC Microbiology, 12*, 280.

Frick, I. M., Schmidtchen, A., & Sjöbring, U. (2003). Interactions between M proteins of Streptococcus pyogenes and glycosaminoglycans promote bacterial adhesion to host cells. *European Journal of Biochemistry, 270*(10), 2303–2311. doi:10.1046/j.1432-1033.2003. 03600.x.

Gerlach, D., Knöll, H., Köhler, W., Ozegowski, J. H., & Hríbalova, V. (1983). Isolation and characterization of erythrogenic toxins. V. Communication: Identity of erythrogenic toxin type B and streptococcal proteinase precursor. *Zentralbl Bakteriol Mikrobiol Hyg A, 255*(2–3), 221–233.

Greenwald, R. J., Freeman, G. J., & Sharpe, A. H. (2005). The B7 family revisited. *Annual Review of Immunology, 23*, 515–548.

Gupta-Malhotra, M. (2004). Antibodies to highly conserved peptide sequence of staphylococcal and streptococcal superantigens in Kawasaki disease. *Experimental and Molecular Pathology, 76*, 117–121.

Haataja, S., & Gerlach, D. (2001). The SpeB virulence factor of Streptococcus pyogenes, a multifunctional secreted and cell surface molecule with strepadhesin, laminin-binding and cysteine protease activity. *Molecular Microbiology, 39*, 512–519.

Haggar, A., Nerlich, A., Kumar, R., Abraham, V. J., Brahmadathan, K. N., Ray, P., et al. (2012). Clinical and microbiologic characteristics of invasive Streptococcus pyogenes infections in north and south India. *Journal of Clinical Microbiology, 50*(5), 1626–1631. doi:10.1128/JCM. 06697-11.

Hedlund, G., Dohlsten, M., Petersson, C., & Kalland, T. (1993). Cancer mmunology mmunoth Papy Superantigen-based tumor therapy: In vivo activation of cytotoxic T cells. *Cancer Immunology, Immunotherapy, 36*, 89–93.

Herwald, H., Cramer, H., Mörgelin, M., Russell, W., Sollenberg, U., Norrby-Teglund, A., et al. (2004). M protein, a classical bacterial virulence determinant, forms complexes with fibrinogen that induce vascular leakage. *Cell, 116*(3), 367–379. doi:10.1016/S0092-8674(04)00057-1.

Hirota, K., Hashimoto, M., Yoshitomi, H., Tanaka, S., Nomura, T., Yamaguchi, T., et al. (2007). T cell self-reactivity forms a cytokine milieu for spontaneous development of IL-17 + Th cells that cause autoimmune arthritis. *The Journal of Experimental Medicine, 204*(1), 41–47. doi:10. 1084/jem.20062259.

Hooker, S., & Follensby, E. (1934). Studies of scarlet fever II. different toxins produced by hemolytic streptococci of scarlatinal origin. *The Journal of Immunology, 27*(2), 177–193.

Ikebe, W. A., Inagaki, Y., Sugama, K., Suzuki, R., Tanaka, D., Tamaru, A., et al. (2002). Dissemination of the phage-associated novel superantigen gene speL in recent invasive and noninvasive Streptococcus pyogenes M3/T3 isolates in Japan. *Infection and Immunity, 70,* 3227–3233.

Ikebe, T., Wada, A., Inagaki, Y., Sugama, K., Suzuki, R., Tanaka, D., et al. (2002). Dissemination of the phage-associated novel superantigen gene spel in recent invasive and noninvasive streptococcus pyogenes M3/T3 isolates in Japan. *Infection and Immunity, 70*(6), 3227–3233. doi:10.1128/IAI.70.6.3227.

Imanishi, K., Igarashi, H., & Uchiyama., T. (1992). Relative abilities of distinct isotypes of human major histocompatibility complex class II molecules to bind streptococcal pyrogenic exotoxin types A and B. *Infection and Immunity, 60,* 5025.

Jing, H. B. (2006). Epidemiological analysis of group A streptococci recovered from patients in China. *Journal of Medical Microbiology, 55,* 1101–1107.

Jolles, S., Sewell, W. A. C., & Misbah, S. A. (2005). Clinical uses of intravenous immunoglobulin. *Clinical and Experimental Immunology, 142*(1), 1–11. doi:10.1111/j.1365-2249.2005.02834.x.

Jones criteria. (1992). Guidelines for the diagnosis of rheumatic fever. 1992 update. *JAMA, 268* (15), 2063–2073.

Kaempfer, R., Arad, G., Levy, R., Hillman, D., Nasie, I., & Rotfogel, Z. (2013). CD28: Direct and critical receptor for superantigen toxins. *Toxins, 5,* 1531–1542. doi:10.3390/toxins5091531.

Kalia, A., & Bessen, D. E. (2003). Presence of streptococcal pyrogenic exotoxin A and C genes in human isolates of group G streptococci. *FEMS Microbiology Letters, 219*(2), 291–295. doi:10. 1016/S0378-1097(03)00022-3.

Kalland, T., Dohlsten, M., Lind, P., Sundstedt, A., Abrahmsen, L., Hedlund, G., et al. (1993). Monoclonal antibodies and superantigens: A novel therapeutic approach. *Medical Oncology & Tumor Pharmacotherapy, 10,* 37–47.

Kalland, T., Hedlund, G., Dohlsten, M., & Lando, P. A. (1991). Staphylococcal enterotoxin-dependent cell-mediated cytotoxicity. *Current Topics in Microbiology and Immunology, 174,* 81–92.

Kamezawa, Y., Nakahara, T., Nakano, S., Abe, Y., Nozaki-renard, J., & Isono, T. (1997). Streptococcal Mitogenic Exotoxin Z, a Novel Acidic Superantigenic Toxin Produced by a T1 Strain of Streptococcus pyogenes. *Infection and Immunity, 65*(9), 3828–3833.

Kapur, V., Nelson, K., Schlievert, P. M., Selander, R. K., & Musser, J. M. (1992). Molecular population genetic-evidence of horizontal spread of 2 alleles of the pyrogenic exotoxin-C gene (spec) among pathogenic clones of streptococcus-pyogenes. *Infection and Immunity, 60*(9), 3513–3517.

Katz, U., Shoenfeld, Y., & Zandman-Goddard, G. (2011). Update on Intravenous Immunoglobulins (IVIg) Mechanisms of Action and Off- Label use in Autoimmune Diseases. *Current Pharmaceutical Design,.* doi:10.2174/138161211798157540.

Kaul, R., Mcgeer, A., Norrby-Teglund, A., Kotb, M., Schwartz, B., Rourke, K. O., et al. (1995). intravenous immunoglobulin therapy for streptococcal toxic shock syndrome—a comparative observational study. *Clinical Infectious Diseases, 28,* 800–807.

Kawabe, Y., & Ochi, A. (1991). Programmed cell death and extrathymic reduction of Vb81 CD41 T cells in mice tolerant to Staphylococcus aureus enterotoxin B. *Nature, 349,* 245–248.

Kotb, M. (1995). Bacterial Pyrogenic Exotoxins as Superantigens. *Clinical Microbiology Reviews, 8*(3), 411–426.

Kotb, M., Norrby-Teglund, A., Mcgeer, A., El-Sherbini, H., Dorak, M. T., Kurshid, A., et al. (2002). An immunogenetic and molecular basis for differences in outcomes of invasive group A streptococcal infections. *Nature Medicine, 8*(12). doi:10.1038/nm.

Kumar, R. K., & Tandon, R. (2013). Rheumatic fever & rheumatic heart disease: The last 50 years. *The Indian Journal of Medical Research, 137*(4), 643–658.

Lamagni, T. L., Darenberg, J., Luca-Harari, B., Siljander, T., Efstratiou, A., Henriques-Normark, B., et al. (2008). Epidemiology of severe Streptococcus pyogenes disease in Europe. *Journal of Clinical Microbiology, 46*(7), 2359–67. doi:10.1128/JCM.00422-08.

Lamagni, T. L., Efstratiou, A., Vuopio-Varkila, J., Jasir, A., & Schalén, C. (2005). The epidemiology of severe Streptococcus pyogenes associated disease in Europe. *Euro Surveillance: Bulletin Européen Sur Les Maladies Transmissibles = European Communicable Disease Bulletin.*

Lancefield, R. C. (1928). The antigenic complex of Streptococcus hemolyticus. I. Demonstration of a type-specific substance in extracts of Streptococcus hemolyticus. *Journal of Experimental Medicine, 47*, 91–103.

Lancefield, R. C., & Dole, V. P. (1946). The properties of T antigen extracted from group A hemolytic streptococci. *Journal of Experimental Medicine, 84*(5), 449–71.

Lando, P. A., Hedlund, G., Dohlsten, M., & Kalland, T. (1991). Bacterial superantigens as anti-tumour agents: induction of tumour cytotoxicity in human lymphocytes by staphylococcal enterotoxin A. *Cancer Immunology, Immunotherapy, 33*, 231–237.

Lebedeva, T., Dustin, M. L., & Sykulev, Y. (2005). ICAM-1 co-stimulates target cells to facilitate antigen presentation. *Current Opinion in Immunology.* doi:10.1016/j.coi.2005.04.008.

LeFebvre, D. (2008). *History of Streptococcus Pyogenes.*

Lei, B., DeLeo, F. R., Reid, S. D., Voyich, J. M., Magoun, L., Liu, M., et al. (2002). Opsonophagocytosis-inhibiting mac protein of group a streptococcus: identification and characteristics of two genetic complexes. *Infection and Immunity, 70*(12), 6880–90.

Leung, D. Y. (1993). A potential role for superantigens in the pathogenesis of psoriasis. *Journal of Investigative Dermatology, 100*, 225–228.

Li, P., Tiedemann, R. E., Moffat, S. L., & Fraser, J. D. (1997). The Superantigen streptococcal pyrogenic exotoxin C (SpeC) exhibits novel mode of action. *Journal of Experimental Medicine, 186*(3), 375–383.

Li, Y., Li, H., Dimasi, N., McCormick, J. K., Martin, R., Schuck, P., et al. (2001). Crystal structure of a superantigen bound to the high-affinity, zinc-dependent site on MHC class II. *Immunity, 14*(1), 93–104. doi:10.1016/S1074-7613(01)00092-9.

Liu, M., Lu, L., Sun, R., Zheng, Y., & Zhang, P. (2015). Rheumatic heart disease: Causes, symptoms, and treatments. *Cell Biochemistry and Biophysics.* doi:10.1007/s12013-015-0552-5.

Llewelyn, M., Sriskandan, S., Peakman, M., Ambrozak, D. R., Douek, D. C., Kwok, W. W., et al. (2004). HLA class II polymorphisms determine responses to bacterial superantigens. *Journal of Immunology (Baltimore, Md.: 1950), 172*(3), 1719–1726. doi:10.4049/jimmunol.172.3.1719.

Madden, J. C., Ruiz, N., & Caparon, M. (2001). Cytolysinmediated translocation (CMT): A functional equivalent of type III secretion in Gram-positive bacteria. *Cell, 104*, 143–152.

Marrack, P., & Kappler, J. (1990). Staphylococcal enterotoxins and their relatives. *Science, 248*, 704–711.

Matsubara, K. (2006). Development of serum IgM antibodies against superantigens of Staphylococcus aureus and Streptococcus pyogenes in Kawasaki disease. *Clinical and Experimental Immunology, 143*, 427–434.

McCormick, J. K., Yarwood, J. M., & Schlievert, P. (2001). Toxic shock syndrome and bacterial superantigens: An update. *Annual Review of Microbiology, 55*, 77–104.

McArthur, J. D., McKay, F. C., Ramachandran, V., Shyam, P., Cork, A. J., Sanderson-Smith, M. L., et al. (2008). Allelic variants of streptokinase from Streptococcus pyogenes display functional differences in plasminogen activation. *The FASEB Journal, 22*(9), 3146–53.

Metzgar, D., & Zampolli, A. (2011). The M protein of group A Streptococcus is a key virulence factor and a clinically relevant strain identification marker. *Virulence, 2*(5), 402–412.

Mollick, J. A., Miller, G. G., Musser, J. M., Cook, R. G., Grossman, D., & Rich, R. R. (1993). A novel superantigen isolated from pathogenic strains of streptococcus pyogenes with aminoterminal homology to staphylococcal enterotoxins B and C. *Journal of Clinical Investigation, 92*(August), 710–719.

Mouthon, L., Kaveri, S., Spalter, S., Lacroix-Desmazes, S., Lefranc, C., Desai, R., & Kazatchkine, M. (1996). Mechanisms of action of intravenous immune globulin in immune-mediated diseases. *Clinical and Experimental Immunology, 104 Suppl*(May), 3–9.

Murakami, J. (2002). Distribution of emm genotypes and superantigen genes of Streptococcus pyogenes isolated in Japan 1994–1999. *Epidemiology and Infection, 128*, 397–404.

Murzin, A. G. (1993). OB(oligonucleotide/oligosaccharide binding)-fold: common structural and functional solution for non-homologous sequences. *The EMBO Journal, 12*(3), 861–867.

Musser, J. M. (1991). Streptococcus pyogenes causing toxic-shock- like syndrome and other invasive diseases: clonal diversity and pyrogenic exotoxin expression. *Proceedings of the National Academy of Sciences, 88*, 2668–2672.

Mylvaganam, H. (2000). Distribution and sequence variations of selected virulence genes among group A streptococcal isolates from western Norway. *APMIS, 108*, 771–778.

Nilsson, M., Sorensen, O. E., Morgelin, M., Weineisen, M., Sjobring, U., & Herwald, H. (2006). Activation of human polymorphonuclear neutrophils by streptolysin O from Streptococcus pyogenes leads to the release of proinflammatory mediators. *Thrombosis and Haemostasis, 95*(6), 982–990. doi:10.1160/TH05-08-0572.

Nobbs, A. H., Lamont, R. J., & Jenkinson, H. F. (2009). Streptococcus adherence and colonization. *Microbiology and Molecular Biology Reviews, 73*, 407–450.

Nomura, Y. (2002). Maternal antibody against toxic shock syndrome toxin-1 may protect infants younger than 6 months of age from developing Kawasaki syndrome. *Journal of Infectious Diseases, 185*, 1677–1680.

Normark, B. H., & Normark, S. (2002). Hosting for the cuel and the inconsequential. *Nature Medicine, 8*(12), 1398–1404. doi:10.1038/nm1202-1350.

Norrby-Teglund, A., Ihendyane, N., Kansal, R., Basma, H., Kotb, M., Andersson, J., & Hammarstro, L. (2000). Relative neutralizing activity in polyspecific IgM, IgA, and IgG preparations against group A streptococcal superantigens. *Clinical Infectious Diseases, 31*, 1175–1182.

Norrby-Teglund, A., Muller, M. P., Mcgeer, A., Gan, B. S., Guru, V., Bohnen, J., et al. (2005). Successful management of severe group A streptococcal soft tissue infections using an aggressive medical regimen including intravenous polyspecific immunoglobulin together with a conservative surgical approach. *Scandinavian Journal of Infectious Diseases, 37*(3), 166–172. doi:10.1080/00365540410020866.

Norrby-Teglund, A., Nepom, G., & Kotb, M. (2002). Differential presentation of group A streptococcal superantigens by HLA class II DQ and DR alleles. *European Journal of Immunology, 32*(9), 2570–2577.

Papageorgiou, A. C., & Acharya, K. (2000). Microbial superantigens: From structure to function. *Trends in Microbiology, 8*, 369–375.

Patel, R., Rouse, M. S., Florez, M. V., Piper, K. E., Cockerill, F. R., Wilson, W. R., & Steckelberg, J. M. (2000). Lack of benefit of intravenous immune globulin in a murine model of group A streptococcal necrotizing fasciitis. *The Journal of Infectious Diseases, 181*(1), 230–234.

Patterson, K. G., Pittaro, J. L. D., Bastedo, P. S., Hess, D. A., Haeryfar, S. M. M., & McCormick, J. K. (2014). Control of established colon cancer xenografts using a novel humanized single chain antibody-streptococcal superantigen fusion protein targeting the 5t4 oncofetal antigen. *PLoS ONE, 9*(4). doi:10.1371/journal.pone.0095200.

Pichichero, M. E., & Casey, J. R. (2007). Systematic review of factors contributing to penicillin treatment failure in Streptococcus pyogenes pharyngitis. *Otolaryngology—Head and Neck Surgery, 137*(6), 851–851.e3.

Ponting, C. P., Marshall, J. M., & Cederholm-Williams, S. (1992). Plasminogen: A structural review. *Blood Coagulation and Fibrinolysis, 3*, 605–614.

Poutsiaka, D. D., Clark, B. D., Vannier, E., & Dinarello, C. A. (1991). Production of interleukin-1 receptor antagonist and interleukin-1 beta by peripheral blood mononuclear cells is differentially regulated. *Blood, 78*(5), 1275–1281.

Prlic, M., & Jameson, S. C. (2002). Homeostatic expansion versus antigen-driven proliferation: Common ends by different means? *Microbes and Infection.* doi:10.1016/S1286-4579(02) 01569-1.

Proft, B. T., Moffatt, S. L., Berkahn, C. J., & Fraser, J. D. (1999). Identification and Characterisation of Novel Superantigens from Streptococcus pyogenes. *Journal of Experimental Medicine, 189*(1), 89–101.

Proft, B. T., Moffatt, S. L., Weller, K. D., Paterson, A., Martin, D., & Fraser, J. D. (2000). Wide Allelic Variation, Mosaic Structure, and Significant Antigenic Variation. *Journal of Experimental Medicine, 191*(10), 1765–1776.

Proft, T., & Fraser, J. D. (2007). Streptococcal superantigens. In *Supernantigens anf Superallergens* (Vol. 93, pp. 1–23).

Proft, T., Webb, P. D., Handley, V., & Fraser, J. D. (2003a). Two novel superantigens found in both group A and group C Streptococcus. *Infection and Immunity, 71*(3), 1361–1369. doi:10. 1128/IAI.71.3.1361.

Proft, T., Yang, L., Fraser, J. D., & Sriskandan, S. (2003b). Superantigens and Streptococcal Toxic Shock Syndrome. *Emerging Infectious Diseases, 9*(10), 1211–1218.

Qian, J., Stenger, B., Wilson, C. A., Lin, J., Jansen, R., Teichmann, S. A., et al. (2001). PartsList: a web-based system for dynamically ranking protein folds based on disparate attributes, including whole-genome expression and interaction information. *Nucleic Acid Research, 29*(8), 1750–1764.

Rapini, R. P., Bolognia, J. L., Jorizzo, J. L. (2007). *Dermatology.*

Reda, K. (1994). Molecular characterization and phylogenetic distribution of the streptococcal superantigen gene (ssa) from Streptococcus pyogenes. *Infection and Immunity, 62*, 1867–1874.

Redpath, S., Alam, S. M., Lin, C. M., O'Rourke, A. M., & Gascoigne, N. R. (1999). Cutting edge: Trimolecular interaction of TCR with MHC class II and bacterial superantigen shows a similar affinity to MHC:peptide ligands. *Journal of Immunology (Baltimore, Md.: 1950), 163*(1), 6–10.

Reid, S. D., Hoe, N. P., Smoot, L. M., & Musser, J. M. (2001). Group A Streptococcus: allelic variation, population genetics, and host-pathogen interactions. *Journal of Clinical Investigation, 107*(4), 393–399. doi:10.1172/JCI11972.

Riley, J. L., & June, C. (2005). The CD28 family: A T-cell rheostat for therapeutic control of T-cell activation. *Blood, 105*, 13–21.

Rodriguez-Iturbe, B., & Musser, J. M. (2008). The current state of poststreptococcal glomeru-lonephritis. *Journal of the American Society of Nephrology, 19*(10), 1855–1864.

Rogers, S., Commons, R., Danchin, M. H., Selvaraj, G., Kelpie, L., Curtis, N., et al. (2007). Strain prevalence, rather than innate virulence potential, is the major factor responsible for an increase in serious group A streptococcus infections. *The Journal of Infectious Diseases, 195*(11), 1625–1633. doi:10.1086/513875.

Roussel, A., Anderson, B. F., Baker, H. M., Fraser, J. D., & Baker, E. N. (1997). Crystal structure of the streptococcal superantigen SPE-C: dimerization and zinc binding suggest a novel mode of interaction with MHC class II molecules. *Nature Structural & Molecular Biology, 4*, 635–643.

Ruiz De Souza, V., Carreno, M. P., Kaveri, S. V., Ledur, A., Sadeghi, H., Cavaillon, J. M., et al. (1995). Selective induction of interleukin-1 receptor antagonist and interleukin-8 in human monocytes by normal polyspecific IgG (intravenous immunoglobulin). *European Journal of Immunology, 25*(5), 1267–1273. doi:10.1002/eji.1830250521.

Schrage, B., Duan, G., Yang, L. P., Fraser, J. D., & Proft, T. (2006). Different preparations of intravenous immunoglobulin vary in their efficacy to neutralize streptococcal superantigens: implications for treatment of streptococcal toxic shock syndrome. *Clinical Infectious Diseases: An Official Publication of the Infectious Diseases Society of America, 43*(6), 743–746. doi:10. 1086/507037.

Schrager H. M., Alberti, S., Cywes, C., Dougherty, G. J., Wessels, M. R. (1998). Hyaluronic acid capsule modulates M protein-mediated adherence and acts as a ligand for attachment of group A streptococcus to CD44 on human keratinocytes. *Journal of Clinical Investigation, 101*, 1708–1716.

Schwartz, J. C., Zhang, X., Fedorov, A. A., Nathenson, S. G., & Almo, S. (2001). Structural basis for co-stimulation by the human CTLA-4/B7-2 complex. *Nature, 410,* 604–608.

Schwartz, R. H. (2003). T Cell Anergy. *Annual Review of Immunology, 21,* 305–334. doi:10.1146/annurev.immunol.20.100301.064807.

Shaikh, N., Leonard, E., Martin, J. M. (2010). Prevalence of streptococcal pharyngitis and streptococcal carriage in children: a meta-analysis. *Pediatrics, 126*(3), e557–64. doi:10.1542/peds.2009-2648.

Sharma, P., Wang, N., & Kranz, D. M. (2014). Soluble T cell receptor Vβ domains engineered for high-affinity binding to staphylococcal or streptococcal superantigens. *Toxins, 6*(2), 556–574. doi:10.3390/toxins6020556.

Sharpe, A. H., & Freeman, G. J. (2002). The B7-CD28 superfamily. *Nature Reviews Immunology, 2*(2), 116–126. doi:10.1038/nri727.

Shiobara, N. (2007). Bacterial superantigens and T cell receptor b-chain-bearing T cells in the immunopathogenesis of ulcerative colitis. *Clinical and Experimental Immunology, 150,* 13–21.

Skansén-Saphir, U., Andersson, J., Björk, L., & Andersson, U. (1994). Lymphokine production induced by streptococcal pyrogenic exotoxin-A is selectively down-regulated by pooled human IgG. *European Journal of Immunology, 24*(4), 916–922. doi:10.1002/eji.1830240420.

Smoot, L. M., et al. (2002). Characterization of two novel pyrogenic toxin superantigens made by an acute rheumatic fever clone of Streptococcus pyogenes associated with multiple disease outbreaks. *Infection and Immunity, 70,* 7095–7104.

Sottini, A., Imberti, L., Gorla, R., Cattaneo, R., & Primi, D. (1991). Restricted expression of T cell receptor Vb but not Va genes in rheumatoid arthritis. *European Journal of Immunology, 21,* 461–466.

Sriskandan, S., Unnikrishnan, M., Krausz, T., Dewchand, H., Noorden, S. Van, Cohen, J., & Altmann, D. M. (2001). Enhanced susceptibility to superantigenassociated streptococcal sepsis in human leukocyte antigen-DQ transgenic mice. *Journal of Infectious Diseases, 184,* 166.

Steer, A. C., Batzloff, M. R., Mulholland, K., & Carapetis, J. R. (2009). Group A streptococcal vaccines: facts versus fantasy. *Current Opinion in Infectious Diseases, 22*(6), 544–552. doi:10.1097/QCO.0b013e328332bbfe.

Steven, D. L. (1992). Invasive group A Streptococcus infections. *Clinical Infectious Diseases, 173,* 619–626.

Stock, A., & Lynn, R. (1969). Extracellular esterases of streptococci and the distribution of specific antibodies in human sera of various age groups. *Journal of Immunology, 102*(4), 859–869.

Sumby, P., Whitney, A. R., Graviss, E. A., Deleo, F. R., & Musser, J. M. (2006). Genome-wide analysis of group a streptococci reveals a mutation that modulates global phenotype and disease specificity. *PLOS Pathogens, 2,* e5.

Sundberg, H., Li, A. S., Llera, J. K., McCormick, J., Tormo, P. M., Schlievert, K., Karjalainen, R. A., & Mariuzza, E. J. (2002). Structures of two streptococcal superantigens bound to TCR-beta chains reveal diversity in the architecture of T cell signaling complexes. *Structure, 10,* 687–699.

Sundberg, E. J., Li, H., Llera, A. S., McCormick, J. K., Tormo, J., Schlievert, P. M., et al. (2002). Structures of two streptococcal superantigens bound to TCR-beta chains reveal diversity in the architecture of T cell signaling complexes. *Structure, 10,* 687–699.

Talkington, D. (1993). Association of phenotypic and genotypic characteristics of invasive Streptococcus pyogenes isolates with clinical components of streptococcal toxic shock syndrome. *Infection and Immunity, 61,* 3369–3374.

Thompson, J. D., Higgins, D. G., & Gibson, T. J. (1994). CLUSTAL W: improving the sensitivity of progressive multiple sequence alignment through sequence weighting, position-specific gap penalties and weight matrix choice. *Nucleic Acids Research, 22*(22), 4673–4680. doi:10.1093/nar/22.22.4673.

Timmer, A. M., et al. (2009). Streptolysin O promotes group A Streptococcus immune evasion by accelerated macrophage apoptosis. *Journal of Biological Chemistry, 284,* 862–871.

Torres, B. A., Kominsky, S., Perrin, G. Q., Hobeika, A. C., & Johnson, H. M. (2001). Superantigens: the good, the bad, and the ugly. *Experimental Biology and Medicine (Maywood, N.J.), 226*(3), 164–176.

Travers, J. (1999). Epidermal HLA-DR and the enhancement of cutaneous reactivity to superantigenic toxins in psoriasis. *Journal of Clinical Investigation, 104*, 1181–1189.

Tripathi, A. (2004). Immunoglobulin E to staphylococcal and streptococcal toxins in patients with chronic sinusitis/nasal polyposis. *Laryngoscope, 114*, 1822–1826.

Tripp, T. J., McCormick, J. K., Webb, J. M., & Schlievert, P. M. (2003). The Zinc-Dependent major histocompatibility complex class II binding site of streptococcal pyrogenic exotoxin C is critical for maximal superantigen function and toxic activity. *Infection and Immunity, 71*(3), 1548–1550. doi:10.1128/IAI.71.3.1548-1550.2003.

Turner, C. (2012). Superantigenic activity of emm3 Streptococcus pyogenes is abrogated by a conserved, naturally occurring smeZ mutation. *PLoS One, 7*, e46376.

Turner, C., & Sriskandan, S. (2007). Streptococcus pyogenes under pressure. *Nature Medicine, 13* (8), 909–910

Uchiyama, S., Andreoni, F., Schuepbach, R. A., Nizet, V., & Zinkernagel, A. S. (2012). DNase Sda1 allows invasive M1T1 Group A Streptococcus to prevent TLR9-dependent recognition. *PLoS Pathogens, 8*(6), e1002736.

Weisman, L. E., Cruess, D. F., & Fischer, G. W. (1994). Opsonic activity of commercially available standard intravenous immunoglobulin preparations. *Pediatric Infectious Disease, 13*, 1122–1125.

Wilson, B. (1952). Necrotising Fasciitis. *American Journal of Surgery, 18*(4), 416–431.

Yoshioka, T. (1999). Polyclonal expansion of TCRBV2- and TCRBV6-bearing T cells in patients with Kawasaki disease. *Immunology, 96*, 465–472.

Yoshioka, T. (2003). Relation of streptococcal pyrogenic exotoxin C as a causative superantigen for Kawasaki disease. *Pediatric Research, 53*, 403–410.

Zhang, C., & Kim, S. H. (2000). A comprehensive analysis of the Greek key motifs in protein beta-barrels and beta-sandwiches. *Proteins: Structure, Function and Genetics, 40*(March), 409–419. doi:10.1002/1097-0134(20000815)40:3<409::AID-PROT60>3.0.CO;2-6.

Zingaretti, C., Falugi, F., Nardi-Dei, V., Pietrocola, G., Mariani, M., Liberatori, S., et al. (2010). Streptococcus pyogenes SpyCEP: A chemokine-inactivating protease with unique structural and biochemical features. *FASEB Journal, 24*, 2839–2848.